SINGLE MOLECULE SCIENCE

T0199650

SINGLE MOLECULE SCIENCE

Physical Principles and Models

Dmitrii E. Makarov

CRC Press
Taylor & Francis Group
Boca Raton London New York

CRC Press is an imprint of the
Taylor & Francis Group, an **informa** business

CRC Press
Taylor & Francis Group
6000 Broken Sound Parkway NW, Suite 300
Boca Raton, FL 33487-2742

First issued in paperback 2020

© 2015 by Taylor & Francis Group, LLC
CRC Press is an imprint of Taylor & Francis Group, an Informa business

No claim to original U.S. Government works

ISBN 13 : 978-0-367-57571-7 (pbk)
ISBN 13 : 978-1-4665-5951-6 (hbk)

Visit the Taylor & Francis Web site at
http://www.taylorandfrancis.com

and the CRC Press Web site at
http://www.crcpress.com

Contents

Preface

This book is on chemical physics and chemical kinetics as viewed through the prism of single-molecule measurements. Traditional chemical kinetics describes how (large) amounts of chemicals evolve in time. A single-molecule measurement zooms in on the elementary processes that cause such time evolution. This book provides an introduction to the mathematical tools and physical theories needed in order to understand, explain, and model single-molecule observations. This book will not teach you how to actually perform single molecule measurements (although the physical principles behind some of the more popular single-molecule techniques are discussed here), but a vast body of already existing literature is concerned with the experimental protocols.

This text does not assume any background beyond undergraduate chemistry, physics, and calculus. Every effort was made to keep the presentation self-contained and to derive or justify every result starting from a limited set of fundamentals (such as several simple models of molecular dynamics and the laws of probability), although the mathematical rigor of some of the proofs may not be up to a purist's standards. The essential concepts used throughout the book (such as the Boltzmann distribution or the rules for working with probabilities) are explained in the two Appendices. Tedious derivations, topics requiring more advanced math, as well as the discussion of issues that are tangential to the main subject are relegated to the sections designated as "Further Discussion." Reading the Further Discussion material is not necessary in order to follow the main body of the book except that some of the mathematical formulas would have to be taken for granted.

The selection of the topics is somewhat geared toward the single-molecule methods used by biophysicists. Within such a narrower scope, the coverage of the relevant models and theoretical ideas is extensive. But given that single-molecule techniques are now widely used across many fields of science, overall cohesion took priority over the book's breadth. For example, I had to (reluctantly) leave out any discussion of single-molecule electronics as a very different set of theoretical ideas and methods would have to be introduced to describe it. Furthermore, of a variety of alternative theoretical approaches to single-molecule phenomena, the discussion is limited to the ones that I felt provided the most intuitive and the least mathematically demanding picture. As a result, some of the topics belonging to the standard repertoire of a theoretical chemical physicist (such as the Fokker-Planck equation) were not included in the book. Finally, some topics were left for the reader to explore as the Exercises randomly dispersed throughout the text.

I am fortunate to have many colleagues and collaborators who have shared their insights with me. Jin Wang was the first person to introduce me to the then emerging field of single molecules about two decades ago and Maria Topaler was my first collaborator on this topic. As a member of Horia Metiu's research group I was lucky to work on a range of problems, from single-photon statistics to kinetic models of protein folding and unfolding, which shaped up my current understanding of those phenomena and got me interested in the field of biophysics. Kevin Plaxco's work on protein folding was another major influence which convinced me to focus on biophysics problems (and to subsequently collaborate with Kevin on a variety of topics).

Benjamin Schuler, besides being a wonderful collaborator, took a particular interest in this book project and provided detailed comments and invaluable advice on almost every chapter. My thinking on many of the topics described in this book was further influenced by many discussions with colleagues, particularly with Ioan Andricioaei, Stanislav Avdoshenko, Paul Barbara, Christopher Bielawski, Johnathan Brantley, Frank Brown, Alan Campion, Srabanti Chaudhury, Ryan Cheng, Yuri Dahnovsky, Atanu Das, Giovanni Dietler, Olga Dudko, William Eaton, Ron Elber, Kilho Eom, E.L. Florin, Irina Gopich, Alexander Grosberg, Helen Hansma, Paul Hansma, Alexander Hawk, Graeme Henkelman, Wilson Ho, Lei Huang, Gerhard Hummer, Tamiki Kamatsuzaki, Serdal Kirmizialtin, Anatoly Kolomeisky, Sai Konda, Christy Landes, Ronald Levy, Hongbin Li, Pai-Chi Li, Andreas Matouschek, Liviu Movileanu, Mauro Mugnai, Abraham Nitzan, Henri Orland, Garegin Papoian, Baron Peters, Steve Presse, Gregory Rodin, Peter Rossky, Michael Rubinstein, Omar Saleh, Reza Soheilifard, Andrea Soranno, John Stanton, John Straub, Attila Szabo, Devarajan Thirumalai, Brian Todd, David vanden Bout, Eric vanden Eijnden, Arthur Voter, Zhisong Wang, X. Sunney Xie, Haw Yang, and Giovanni Zocchi. Of course, I am solely responsible for any misinformation that may result from reading in this book.

1 A Brief History of Thought and Real Single-Molecule Experiments

> No one has ever seen or handled a single molecule. Molecular science, therefore, is one of those branches of study which deal with things invisible and imperceptible by our senses, and which cannot be subjected to direct experiment.
>
> James Clerk Maxwell, *Molecules*

At the time of writing, the existence of molecules has been viewed as proven for about a century. Although various philosophers had argued that matter must consist of small, indivisible particles for millennia, early scientific evidence for molecules came around the early 19th century after chemists noticed that, in order to form chemical compounds, their components or "elements" have to be mixed in certain simple proportions. For example, carbon was known to form two different oxides. To make the first one out of one gram of carbon, it would have to be combined with 1.333 grams of oxygen. In contrast, the second oxide would require 2.666 grams of oxygen. John Dalton explained such observations by positing that chemical compounds consist of molecules, which, in turn, are formed from atoms. The molecule of the first oxide (CO, in modern notation) consists of one carbon atom and one oxygen atom, while the second one (CO_2) has one carbon and two oxygen atoms, which immediately explains why it requires twice as much oxygen. Furthermore, one readily concludes that the ratio of the mass of the oxygen and the carbon atom must be equal to $1.333 \approx 4/3$. By analyzing such proportions in various chemical compounds, an internally consistent list of *relative* atomic masses was established. For example, if the hydrogen mass is taken as the atomic mass unit, then the masses of carbon, nitrogen, and oxygen are, respectively, equal to 12, 14, and 16. Unfortunately, the *absolute* mass of an atom cannot be determined in this way. Further refined by his followers, notably by Amedeo Avogadro who proposed the existence of diatomic molecules such as O_2 or H_2, Dalton's ideas became widely accepted as the 19th century progressed. Despite overwhelming indirect evidence for Dalton's atomic theory, however, the lack of any direct observations of atoms and molecules led many prominent scientists to view molecules as convenient mathematical devices rather than real physical entities.

Remarkably, estimates of physical properties of molecules such as their mass, speed, or size had been deduced from experimental observations long before molecules could be observed. Those estimates were based on the bold proposition that macroscopic properties of gases originate from random motion of their constituent molecules. First put forth around 1738 by Daniel Bernoulli and currently known as "kinetic theory of gases," this theory, in particular, explains the pressure that a gas exerts on the walls of its container as a result of incessant bombardment by the gas molecules. Let u_x be the component of the molecule's velocity measured, at the instant just before the molecule strikes the container wall, along the axis x that is perpendicular to the wall. Assuming a perfectly elastic collision, u_x changes its sign once the molecule bounces off the wall. As a result, a momentum $mu_x - (-mu_x) = 2mu_x$, where m is the molecule mass, is transferred to the wall. According to Newton's second law, the total force exerted by the gas on the wall equals the momentum transferred to the wall by all the molecules per unit time. The molecules that strike the wall during a short time interval Δt are those that happened to be close enough to reach the wall. Specifically, they must be within the distance $u_x \Delta t$ from the wall. The number of such molecules is $\rho A u_x \Delta t / 2$, where ρ is the number of molecules per unit volume, A is the area of the wall, and the factor $1/2$ accounts for the fact that half of the molecules contained within the volume $A u_x \Delta t$ are actually moving away from the wall and should not be counted. The number of molecules that hit the wall, per unit time, is thus $\rho A u_x / 2$, resulting in a force that is normal to the wall and equal to $2mu_x \rho A u_x / 2 = m\rho A u_x^2$. Since the velocities of different molecules are different, this result has to be averaged over all the molecules. Using angular brackets to denote an average quantity and recognizing that the gas pressure P is the force per unit area, we find

$$P = \rho m \langle u_x^2 \rangle = \frac{1}{3} \rho m \langle u^2 \rangle.$$

Here u denotes the total velocity of a molecule and the following obvious identity is used,

$$\langle u^2 \rangle = \langle u_x^2 + u_y^2 + u_z^2 \rangle = 3\langle u_x^2 \rangle,$$

with u_y and u_z being the components of the velocity along two axes orthogonal to x. Finally, notice that

$$\rho_m = \rho m$$

is the mass density (i.e., the mass per unit volume) of the gas. Therefore, all one needs to know in order to estimate the root mean square velocity of a molecule is the gas pressure and its mass density:

$$\langle u^2 \rangle^{1/2} = \sqrt{3P/\rho_m}.$$

The result depends on the gas in question and, typically, ranges from hundreds to thousands meters per second. To a physicist of the 19th century, it must have been shocking to conclude that his skin is bombarded by little projectiles each moving as fast as a bullet from a gun.

But what perhaps was even more surprising was that those projectiles did not travel very far on the average: collisions with other molecules cause them to frequently change their course, with the resulting net motion being very slow. In a lecture reported in a 1873 issue of *Nature* [1], James Clerk Maxwell opened a bottle of ammonia. If each ammonia molecule traveled along a straight line, it would have reached the walls of the lecture hall in a fraction of a second. Yet, as a result of collisions and the ensuing erratic motion, it took an appreciable time before the audience could smell the chemical. A more quantitative version of Maxwell's lecture demonstration enabled another 19th century physicist, Johann Josef Loschmidt, to estimate the size of a molecule, from which the number ρ of molecules per unit volume (often referred to as the Loschmidt number) and, consequently, the molecular mass could further be deduced. Loschmidt's experiments with gas diffusion allowed him to estimate the mean free path, i.e., the average distance λ a molecule travels before colliding with another molecule. Without going into details of the actual measurements, let us show how λ can, in principle, be estimated from Maxwell's demonstration.[1] The trajectory of a selected ammonia molecule consists of connected straight-line segments, each having a random direction. Although all the segments do not have to be of the same length, an order-of-magnitude estimate of the mean segment length λ can be obtained if we pretend that all of them have a length equal exactly to λ. Mathematical properties of such trajectories, also referred to as random walks, are discussed in Appendix A. Using the results from the appendix, the mean square distance traveled by the ammonia molecule away from the bottle is given by

$$\langle r^2 \rangle = n\lambda^2,$$

where n is the number of steps in the walk (i.e., the number of segments), which can be estimated as the length of the trajectory of a molecule traveling during a time t with a typical velocity u (which we have already estimated) divided by the length of each straight segment:

$$n \approx ut/\lambda.$$

Thus we have

$$\langle r^2 \rangle \approx \lambda ut.$$

If r is taken to be the length of the hall where Maxwell lectured, then

$$t = \frac{r^2}{\lambda u}$$

can be taken as a crude estimate of the time it takes the ammonia smell to spread uniformly over the room. Of course, some of the molecules may travel the distance r and reach the listeners' noses sooner than the mean time t predicts so a more careful analysis would require consideration of the probability distribution of r derived in Appendix A. Given our unrealistic assumption of perfectly still air, however, such more refined estimates will not be pursued here. The mean free path can now be estimated, once the size of the lecture hall r and the smell spreading time t are measured, as

$$\lambda \approx \frac{r^2}{ut}.$$

It is clear that the mean free path must be related to the molecular size d for, if the molecules were infinitely small, they would never collide with one another. This relationship can be crudely estimated if we think of all molecules as hard spheres of diameter d. Imagine tracking the path of a selected molecule. During some time interval t, it will collide with every molecule whose center came within the distance d from the molecule's path. Since the length of the path is $l = ut$ then the number of the molecules that act as obstacles is simply $\pi d^2 l \rho$, where, again, ρ is the number

---

[1] In practice, our estimate would not be very realistic as it requires the air in the lecture hall to be perfectly still.

of molecules per unit volume of the gas. The average distance traveled between two collisions is, therefore,

$$\lambda = \frac{l}{\pi d^2 l \rho} = \frac{1}{\pi d^2 \rho}.$$ (1.1)

If one of the two quantities, ρ or d, is known, then the other one can be calculated from Eq. 1.1. However the experimental information considered so far does not appear to allow independent estimation of either of the two. Loschmidt realized that yet enough relationship between d and ρ can be established by comparing the volume occupied by a gas, V_g, and the volume of the same material in the liquid form, V_l. Suppose the material contains N molecules. Then, by definition, we have

$$\rho = \frac{N}{V_g}.$$ (1.2)

On the other hand, Loschmidt reasoned, the typical distance between the molecules of a liquid must be comparable to the molecular size d. Indeed, close proximity of molecules is supported by the fact that liquids are nearly incompressible. The volume of the liquid then can be estimated as

$$V_l \approx cNd^3.$$ (1.3)

Even if we believe that molecules are truly spheres, the exact numerical proportionality coefficient c is not easy to calculate unless the molecules are packed in an orderly fashion (which we know is not the case for a liquid). For an order-of-magnitude estimate, this numerical factor will simply be omitted. For consistency, we will also drop π from Eq. 1.1. Taking the ratio of Eqs.1.2 and Eqs.1.3, we now arrive at the sought after independent relationship between ρ and d:

$$\rho d^3 \approx \frac{V_l}{V_g}.$$ (1.4)

Rewriting Eq. 1.1 as $\lambda \approx \frac{d}{d^3\rho}$ and using Eq. 1.4, we find

$$\lambda \approx d\frac{V_g}{V_l}.$$

This gives an estimate of the molecular size,

$$d \approx \lambda\frac{V_l}{V_g},$$

which, to within a numerical factor, agrees with that of Loschmidt. Specifically, using experimental data on gas diffusion, he estimated d to be in a nanometer range, in remarkably good agreement with our modern knowledge. The number of molecules per unit volume can now be estimated using Eq. 1.1, which gives:

$$\rho \approx \frac{1}{\lambda d^2} \approx \frac{1}{\lambda^3}\left(\frac{V_g}{V_l}\right)^2.$$

Maxwell's estimate for this number, which is now called the Loschmidt constant, is $1.9 \times 10^{25} m^{-3}$ [1]. This estimate is remarkably close to the modern value of $\approx 2.69 \times 10^{25} m^{-3}$ at $0°C$ and one atmosphere.

A different piece of evidence for molecules as building blocks of all materials came from the discovery of incessant random motion (Brownian motion) exhibited by small material particles suspended in water, which is usually attributed to the Scottish botanist Robert Brown. In 1905–1906, Albert Einstein and, independently, Marian Smoluchowski developed quantitative theories of Brownian motion based on the premise that it originates from the bombardment of the Brownian particles by the surrounding molecules of water. Some of the predictions of their theories will be described in Chapter 4 of this book. In particular, the Einstein-Smoluchowski equation, Eq. 4.17, relates the viscous drag experienced by a Brownian particle and its diffusion coefficient, which both can be measured experimentally. This equation contains the physical constant k_B that relates the energy of a molecule to temperature and is known as the Boltzmann constant.[2] As a result, the value of the Boltzmann constant k_B can be estimated. On the other hand, kinetic theory of gases predicts the Boltzmann constant to be related to the universal gas constant R, which is the experimentally measured proportionality constant in the ideal gas equation of state,

$$P V_m = RT,$$

where V_m is the volume occupied by one mole of an ideal gas at pressure P and temperature T. The relationship between the microscopic constant k_B and the macroscopic constant R is given by

$$k_B = R/N_a,$$

where N_a is the number of molecules in one mole, referred to as Avogadro's number. Once k_B is known, Avogadro's number can also be estimated.[3] Finally, the mass of a molecule can now be estimated by dividing the mass of one mole by Avogadro's number. Experiments conducted by Jean Baptiste Perrin in 1908 confirmed the predictions of the theory of Brownian motion and yielded estimates of k_B, N_a and other microscopic properties. Despite the lack of direct observation of molecules, those experiments were generally viewed as the final proof that molecules are real.

The preceding discussion attempted to describe a century or so of efforts in gathering and interpreting circumstantial evidence for molecules and atoms. But when was a molecule first observed? In response to this question, my colleagues gave diverse answers. Should we count jewelers as single-molecule experimenters given that, technically speaking, a diamond crystal is a single molecule? Probably not. The detection of microscopic (particularly, radioactive) particles has been demonstrated as early as in the beginning of the 20th century, when Hans Geiger, Ernst Marsden, and Ernest Rutherford performed their famous experiments with alpha particles and when C.T.R.

---

[2] For example, the mean translational kinetic energy of a molecule at a temperature T is $3k_B T/2$. See Appendix B for further details.

[3] Note that Avogadro's number is related to the Loschmidt constant through $\rho = \frac{N_a P}{RT}$ and so Loschmidt's estimate should be regarded as the first estimate of N_a as well. Also note that Max Planck independently estimated the value of k_B from his theory of black body radiation, which was published in 1900 and which is regarded as the beginning of the quantum theory.

Wilson invented the cloud chamber. Notwithstanding those early advances, the birth of single-atom and single-molecule measurements is usually attributed to several developments that occurred in the 1970–1980s. In the field of electrophysiology, reversible jumps in the conductance of cell membranes were interpreted as arising from the conformational changes in single ion channels [4,5], i.e., pore-forming membrane proteins that can selectively control the passage of ions through cell membranes. In a distinctly unrelated field of quantum optics, Kimble et al. observed fluorescence photons emitted by individual atoms in an atomic beam [2,3]. Their experiment analyzed temporal correlations among photons emitted at different times and confirmed the "photon antibunching" effect predicted earlier by quantum theory: unlike the photons emitted by multi-atom sources, photons from a single atom tend to "repel" one another so that one photon is unlikely to be immediately followed by the next. The photon antibunching effect, which will be further discussed in Chapter 7 of this book, is now routinely used to verify whether observed light has a single-molecule origin.

The experiments of Kimble et al. employed an atomic beam but were performed under the conditions such that no more than a single atom was typically in the observation volume at any moment. Later studies showed that individual atomic ions could be confined within electromagnetic traps for a long time [6]. Those experiments employed radiation pressure from tunable lasers to slow down the motion of atoms thereby cooling them to extremely low temperatures at which they could not escape the applied electromagnetic field. In contrast to atoms, however, cooling molecules with lasers remained elusive because, in addition to their translational motion, molecules also undergo rotation. This problem was circumvented by a different approach, in which the molecules of interest were trapped within a solid [7,8]. Achieving a single-molecule level of resolution within this technique was facilitated by a phenomenon that is normally considered a nuisance: because of the interaction of the "guest" molecule with a "host" solid, each molecule sees a somewhat different local environment and, as a result, absorbs and emits light of somewhat different wavelength. This results in the so-called inhomogeneous broadening effect, where the observed spectral lines are much broader than those expected from an individual molecule. But what was considered to be a handicap in the bulk spectroscopy of solids turned out to be a blessing in disguise for the single-molecule spectroscopists: when the concentration of guest molecules in the solid was low enough the laser light wavelength could be tuned so as to selectively excite only one guest molecule within the observation volume while the surrounding guest molecules remained "dark."

Other important developments that occurred in the 1980s included the invention of the scanning tunneling microscope (STM) and the atomic force microscope (AFM), which allowed imaging and manipulation of individual molecules at surfaces. While early single-molecule spectroscopy studies required low temperatures, this limitation was circumvented in the 1990s (see ref. [9] for a review), paving the way for single-molecule studies of biochemical processes and living systems. As a result, the single molecule field saw explosive development in the 1990s, when numerous new experimental methods were developed and used in chemistry, physics, molecular biology, and materials science. By now, single-molecule experimental techniques have evolved from being technological marvels to nearly routine tools, although many exciting developments, particularly concerning the improvement in temporal and spatial resolution, are still underway.

Finally, it should be noted that the behavior of systems containing one or few molecules has also been the subject of considerable theoretical thought, which often preceded experimental observations. In the early 1960s, for example, Terrell L. Hill published a book [10] that anticipated the future need for extending thermodynamics—which is conventionally formulated for systems of infinite size—to microscopic objects. The advent of computer simulations in the second half of the 20th century has further led to the development of Molecular Dynamics and Monte Carlo methods, which track the fate of individual molecules on a computer. Although the original purpose of those studies was not necessarily to observe the behavior of individual molecules, virtual single-molecule computer experiments became a natural by-product. As the reader will see from the following chapters of this book, recent discoveries made through single-molecule experiments continue to go hand in hand with theoretical advances in molecular science.

REFERENCES

1. James Clerk Maxwell, "Molecules," *Nature*, vol. 8, 437-441, 1873.
2. H.J. Kimble, M. Dagenais, and L. Mandel, "Photon antibunching in resonance fluorescence," *Phys. Rev. Letters*, vol. 39, 691, 1977.
3. D.F. Walls, "Evidence for the quantum nature of light," *Nature*, vol. 280, 451-454, 1979.
4. G. Ehrenstein, H. Legar, and R. Nossal, "The nature of the negative resistance in bimolecular lipid membranes containing excitability-inducing material", *The Journal of General Physiology*, vol. 55, 119-133, 1970.
5. E. Neher, B. Sakmann, "Single-channel currents recorded from membrane of denervated frog muscle fibres", *Nature*, vol. 260, 799-802, 1976.
6. Wayne M. Itano, J. C. Bergquist, and D. J. Wineland, "Laser spectroscopy of trapped atomic ions", *Science*, vol. 237, 612, 1987.
7. W.E. Moerner and L. Kador, "Optical detection and spectroscopy of single molecules in a solid", *Phys. Rev. Letters*, vol. 62, 2535, 1989.
8. M. Orritt and J. Bernard, "Single pentacene molecules detected by fluorescence excitation in a p-terphenyl crystal", *Phys. Rev. Letters*, vol. 65, 2716, 1990.
9. X. Sunney Xie, "Single-molecule spectroscopy and dynamics at room temperature", *Acc. Chem. Res.*, vol. 29, 598, 1996.
10. Terrell L. Hill, *Thermodynamics of Small Systems*, Dover Publications, 1994.

2 How the Properties of Individual Molecules Are Measured

... Why are atoms so small? Clearly, the question is an evasion. For it is not really aimed at the size of the atoms. It is concerned with the size of organisms, more particularly with the size of our own corporeal selves. Indeed, the atom is small, when referred to our civic unit of length, say the yard or the metre.

Erwin Schrödinger, *What is Life?*

2.1 TYPICAL SIZE OF A MOLECULE

The distance between the oxygen atoms in the O_2 molecule is about $R_{O-O} \approx 1.2 \times 10^{-10}$ meters. One Angström (denoted $1 \mathring{A}$) is equal to 10^{-10} meters and one nanometer (1 nm) is 10^{-9} meters. The distance R_{O-O} is thus equal to $1.2\mathring{A} = 0.12$ nm. To appreciate how small this length is, imagine drawing this molecule on your notepad as two circles (representing the oxygen atoms) connected by a line. If the distance between the circles is, say, 10 centimeters, then it represents the oxygen molecule magnified by about billion (10^9) times. For comparison, the size of your drawing magnified by another billion times would be comparable to the distance from the Earth to the Moon.

Some molecules are considerably larger than the simple diatomics such as O_2. The DNA that encodes the human genome, for example, is a molecular chain that, if fully stretched, would have a length of a few centimeters. Many polyatomic molecules fall somewhere in between these two extremes. In particular, many molecules of life have linear dimensions of a few nanometers. We will consider this to be the "typical" molecular size.

The tremendous disparity between microscopic, molecular length scales and the typical dimensions of the macroscopic devices we use to make measurements presents a tremendous hurdle to single-molecule observation. Interestingly, human sensory systems can, in certain cases, successfully overcome this hurdle and provide us with an interface to the microscopic world: For example, our eye is capable of detecting individual photons![1]

---

[1] Our brain, however, is designed to ignore single photons and sets the detection threshold to several photons

2.2   OPTICAL DETECTION OF AN INDIVIDUAL MOLECULE

Our eye analyzes the wavelength and the intensity distribution of the light emitted, reflected, or scattered by an object thereby providing us information about its shape and color. Typically, the light we see originates from an astronomic number of molecules. Shapes and colors of small objects that cannot be resolved with the naked eye are conveniently examined with a magnifying glass or a microscope. Could we then simply use a very powerful microscope to look at the structure of a single molecule?

Unfortunately, the wave nature of light imposes a fundamental limit on the resolution of an optical microscope. Specifically, light diffraction causes any small object observed through a microscope to appear blurred. Consequently, spatial details of an object can be discerned only if the length scale of such details is greater than, roughly, the wavelength λ of light used to observe the object. The wavelength of visible light is in the range 400–700 nm, orders of magnitude greater than a typical molecular size. Therefore, visible light scattered or emitted by a molecule carries little information about its geometry or structure. Nevertheless, this light can be used to detect *the presence* of a molecule in a spot whose size is comparable to λ.[2] This approach could be promising if we could come up with a sample containing 1 molecule per λ^3, or, equivalently,

$$\frac{1}{\lambda^3} \approx \frac{1}{(500\text{nm})^3} \approx 10^{19} \frac{\text{molecules}}{m^3}.$$

For comparison, air contains approximately 2.5×10^{25} molecules per cubic meter while liquids or solids typically contain roughly 10^{28} molecules per cubic meter. It then appears that observing a single molecule in a condensed phase is impossible, as many other molecules are bound to fall within the same observation spot and interfere with the image.

Fortunately, there is a solution to this problem. Molecules can either absorb or emit light. When light is absorbed, the energy of a photon, $h\nu$, is utilized to drive the molecule from a low energy state E_1 to a high energy state E_2,

$$E_2 = E_1 + h\nu, \tag{2.1}$$

where $\nu = c/\lambda$ is the frequency of light, c is the speed of light, and the proportionality coefficient between the photon energy and its frequency, $h \approx 6.626 \times 10^{-34} J \times s$, is known as Planck's constant. Likewise, a transition from a higher energy level E_2 to a lower level E_1 may be accomplished by photon emission. The wavelength of the emitted or absorbed light, therefore, depends on the energy levels $E_1, E_2, \ldots$ of the molecule which, in turn, depend on its chemical identity. Different molecules have different light absorption/emission wavelengths. Textile industry routinely takes advantage of this fact: By incorporating various "dye" molecules within, e.g., cotton fibers, colorful t-shirts can be readily produced. Likewise, chemists take advantage of different "colors" of the molecules to distinguish among them. The scientific term

---

[2] Moreover, the location of the molecule within such a spot can be determined with an accuracy that is better than λ using the so-called superresolution methods. The idea behind superresolution is that a diffraction-blurred image of a molecule is nonuniform and, moreover, is centered around the molecule's true location

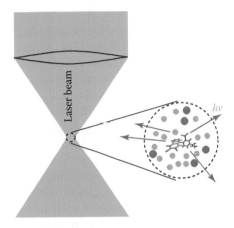

FIGURE 2.1 Single-molecule fluorescence spectroscopy experiment: A laser beam is focused in a tight spot, typically of about a micrometer in size. The wavelength of light is chosen so as to excite the fluorescent molecule of interest and the photons emitted by this molecule are detected with a single-photon detector (not shown). The concentration of the fluorescent molecules is so low that, typically, no more than one such molecule is found within the observation spot. This molecule may be surrounded by many other chemical species but those do not fluoresce (in the wavelength range probed) and so they do not impede the observation of the fluorescent molecules under study.

for this method is *spectroscopy*. A typical single-molecule spectroscopy setup is illustrated in Figure 2.1 and involves a laser beam focused into a small spot. The minimum possible size of the spot is, again, determined by light diffraction and, crudely, is comparable with the wavelength λ. Thus we estimate the volume of the spot to be

$$V \approx \lambda^3 \approx (0.1 - 1) \times 10^{-18} m^3 = (0.1 - 1) \text{ femtoliters.}$$

The wavelength of laser light is chosen so as to drive certain kind of molecules (that we want to study) to a higher energy state. As a result, those molecules can re-emit light through the process called fluorescence. By properly tuning the wavelength of light, the molecules of interest can be driven to emit light while the surrounding molecules remain dark because they have no energy levels satisfying Eq. 2.1. To observe the fluorescence light emitted by a single molecule, all we have to do now is to ensure that the concentration of the fluorescent molecules is low enough that no more than one of them is located within the detection volume V. This can be achieved, e.g., by preparing a very dilute solution of the fluorescent molecules, which contains fewer than $1/V \approx 10^{18} - 10^{19}$ molecules per cubic meter. Chemists prefer to measure concentrations of molecules in moles per liter and use the capital letter "M" to indicate these units. One mole per liter corresponds to Avogadro's number of molecules contained in one liter of solution,

$$1M \approx 6.02 \times 10^{23}/10^{-3} m^{-3} = 6.02 \times 10^{26} m^{-3}.$$

In terms of these chemical units, the fluorescent molecules must have a concentration of less than 1–10 nM. It should be empasized that the total number of molecules in the detection volume V is unimportant as long as most of them do not emit light in response to the laser excitation. While it was implied that the fluorescent molecules were in a liquid, they can also be embedded within a solid or immobilized on a solid surface. Because such immobilization may affect the properties of the molecules, studies of immobilized single molecules are more intrusive than liquid phase experiments.

Experiments with unconstrained molecules in solution, however, entail a different complication: How can we keep the molecule within the very small detection volume V? Typical thermal velocities u of molecules range from tens to hundreds of meters per second, depending on their mass. At first glance it appears that the time such a molecule would remain within the detection volume, whose linear size is comparable to the wavelength λ, would be λ/u, which is in a nanosecond range. In reality, the typical dwell time is orders of magnitude longer, typically in a millisecond range. This lucky break is due to the help from the numerous surrounding molecules present within the detection volume: Our fluorescent molecule bumps into its neighbors and changes its direction of motion many times before it escapes. Think of trying to escape a crowded mall: Bumping into other people slows you down. Some of the mathematical properties of the resulting diffusive motion will be discussed later in this book.

The picture of a solution phase experiment is therefore as follows: A fluorescent molecule arrives at the detection volume created by the laser focal spot, an event signaled by a burst of photons arriving at the photodetector. The experimenter has about a millisecond or so to collect those photons before the molecule exits the detection volume. Then he or she has to wait for the next one to enter the focal spot. But what are such observations good for? Undoubtedly, they would inform us that a molecule has visited the focal spot of the laser beam but so what? What we really want is to learn something about the properties of the molecule itself. It turns out that the properties of the observed photons, particularly the color (i.e. the wavelength) and their arrival times, can provide us plenty of information about the molecule's structure and internal motion because the color can be correlated with the molecule's configuration. How to decipher this information from the experimental signal will be the subject of Chapter 7.

2.3   SCANNING PROBE MICROSCOPIES

A family of experimental techniques called scanning probe microscopies offers another way to observe and even manipulate individual molecules. The basic idea behind those methods is to drag a very small probe along a surface, on which the molecules of interest are placed (Fig. 2.2). The probe is usually shaped as a very sharp tip. Use of piezoelectric actuators allows the experimenter to control the tip position with subnanometer precision. As the tip is moved along the surface, its interaction with the surface atoms results in a measurable experimental signal, as a function of the tip position. For example, in the Atomic Force Microscope (AFM) the tip is attached to a cantilever (Fig. 2.2), which is deflected as a result of the interactions between the tip and the surrounding molecules. This allows one to measure the force exerted by the nearby molecules on the tip. By moving the tip along the surface, it is possible to image the molecules lying on the surface. For example, AFM images of DNA are shown in Figure 2.3. Likewise, a scanning tunneling microscope (STM) measures the

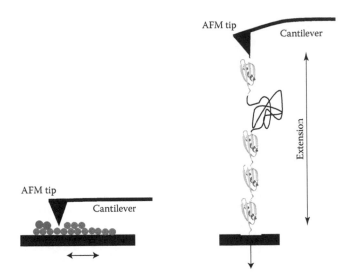

FIGURE 2.2 Atomic force microscope (AFM) can be used to measure the force exerted on a tip by molecules adsorbed at a surface and thus image the surface "topography" (left). AFM can also be used to stretch long chain molecules mechanically (right). Note that the size of the AFM tip is not to scale here: in reality it is much larger relative to the size of the molecules probed.

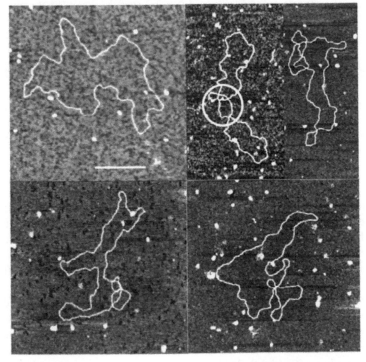

FIGURE 2.3 An AFM image of DNA molecules can resolve such structural details as knots in the DNA chain. Reprinted with permission from E. Ercolini et al., *Phys. Rev. Letters*, 98 058102 (2007). Copyright (2007) by the American Physical Society.

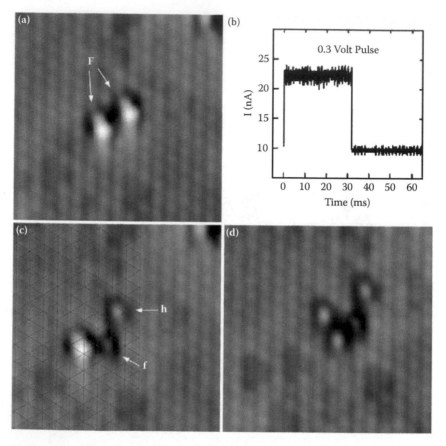

FIGURE 2.4 (a) An STM image of two oxygen molecules on a platinum surface. Upon application of a voltage pulse (b) to one of the molecules, it is seen to dissociate into a pair of oxygen atoms (c). The second molecule dissociates (d) upon application of another voltage pulse. Reprinted with permission from B.C. Stipe et al., *Phys. Rev. Letters*, 78 4410 (1997). Copyright (1997) by the American Physical Society.

electric current between the tip and the surface atoms. Figure 2.4 shows an example where oxygen molecules adsorbed on a tungsten surface were imaged using an STM with resolution so high that the orientation of an individual molecule could be discerned in the image.

As immobilization of the molecule of interest on a surface requires a sufficiently strong molecule-surface interaction, scanning probe microscopies tend to strongly perturb the properties of the molecules under study. While this is a downside when the properties of free, isolated molecules are desired, the interaction of molecules with the experimental setup can also be used to the experimenter's advantage. For example, AFM can be used to exert mechanical pulling forces on molecules, as illustrated in Figure 2.2 (right). This scenario will be further explored in Chapters 8 and 9. Likewise, strong electric interactions between the STM tip and a molecule can be used to manipulate this molecule. For example, application of a voltage between

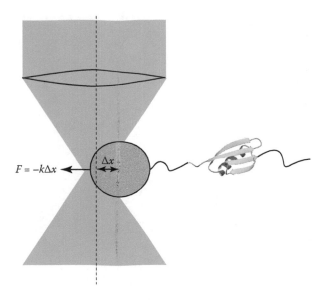

$F = -k\Delta x$

Δx

FIGURE 2.5 Optical tweezers: Interaction of a dielectric bead with light from a laser can be used to exert mechanical forces on molecules. Note that, in reality, the bead would be much larger than the protein molecule depicted here.

the surface and the tip may cause the O_2 molecule to fall apart, as illustrated in Figure 2.4.

2.4 OPTICAL TWEEZERS

Optical tweezers offer another way to exert mechanical forces on individual molecules. Optical tweezers utilize the interaction of a micrometer-sized bead with a tightly focused laser beam to exert mechanical force on a molecule that is attached to the bead. In a setup shown in Figure 2.5, this interaction creates a restoring force that drives the bead towards the middle of the beam. The typical forces generated by tweezers are of order of tens of piconewtons or less. It turns out that those are comparable to the typical forces exerted on (and by) biomolecules in living organisms. For this reason, optical tweezers are particularly well suited for studying mechanical phenomena occurring in the living cell. We will have a further discussion of optical tweezers experiments in Chapters 8–10.

2.5 NANOPORE EXPERIMENTS

Another single-molecule method, which is quite different from all the other approaches described so far, employs the passage of molecules through small, nanometer-sized pores. Such pores are often conveniently provided by biological systems: For example, the α-hemolysin pore shown in Figure 2.6 is an assembly of proteins that is produced by *Staphylococcus aureus*, a bacterium that is responsible for several

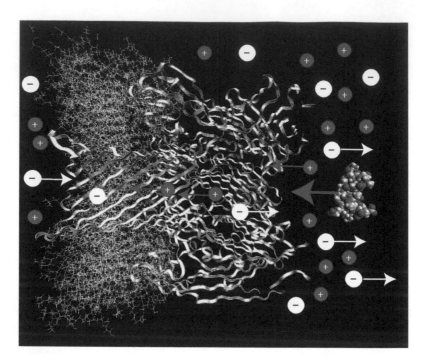

FIGURE 2.6 In nanopore experiments, an electric field drives both solution ions and molecules of interest (e.g., DNA or peptides) across nanometer-sized pores. The specific pore depicted here (and commonly employed in nanopore studies) is the α-hemolysin pore formed by several protein units that are embedded within a lipid bilayer (image courtesy of Kijeong Kwac). Depending on its structure and conformation, a bulky molecule entering the pore may completely or partially block the ionic current across the pore. The duration and the magnitude of a blockade event enable the experimenter to recognize the molecule's identity and/or structure

common types of infections. Those proteins self-assemble on a cell membrane to form a channel, through which molecules, normally confined within the cell, can leak out, resulting in cell destruction. While detrimental to living cells, the α-hemolysin pore provides a convenient tool for single-molecule experiments, where various electrically charged macromolecules (most commonly, DNA and polypeptides) are driven across this pore by a voltage difference applied across the membrane. When the nanopore is immersed in an ionic solution, the electric field also drives positively and negatively charged ions through the pore. This ionic current across a single nanopore constitutes the observable experimental signal. Bulky molecules that occasionally get dragged into the pore obstruct the passage of the ions thereby reducing the ionic current. The amount by which the current is suppressed, as well as the time the molecule spends inside the pore, depends on the identity and the structure of the molecule inside the pore. This dependence, for example, enables single-molecule DNA sequencing, where chemical identities of individual DNAs passing through the pore are discerned.

Like pulling experiments employing AFM or optical tweezers, nanopore experiments involve exertion of a tunable mechanical force on the molecule of interest. Indeed, this force can be changed by varying the voltage across the pore. Some of the general features of mechanical single-molecule experiments and their implications for the physics of living systems will be further discussed in Chapters 8–10. Finally, we note here that nanopore experiments do not have to employ Nature-made protein pores; recent nanopore studies often take advantage of engineered pores, which are often more robust and versatile than their biological counterparts, particularly in their ability to withstand high voltages and tunability of their size.

The rather limited account of experimental single-molecule methods presented in this chapter does not do justice to the thriving single-molecule field. There are many more ingenuous ideas allowing exploration of various properties of individual molecules. As there are entire books dedicated to the subject of experimental single-molecule techniques, the reader interested in performing actual single-molecule measurements will benefit from further reading on this topic. Refs. [1] and [2] would be good starting points in this direction.

REFERENCES

1. Paul R. Selvin and Taekjip Ha (Editors), *Single-molecule Techniques: A Laboratory Manual*, Cold Spring Harbor Laboratory Press, 2008.
2. Tamiki Komatsuzaki, Masaru Kawakami, Satoshi Takahashi, Haw Yang, and Robert J. Silbey (Editors), *Single Molecule Biophysics: Experiments and Theory*, Wiley, 2012.

3 The Kinetics of Chemical Reactions: Single-Molecule Versus "Bulk" View

The single experience of one coin being spun once has been repeated a hundred and fifty six times. On the whole, doubtful. Or, a spectacular vindication of the principle that each individual coin spun individually is as likely to come down heads as tails and therefore should cause no surprise each individual time it does.

Tom Stoppard, *Rosencrantz and Guildenstern are Dead*

Chemistry is a discipline that studies how molecules convert into one another. A chemical reaction can involve several different molecules. For example, a molecule of sodium chloride can fall apart (i.e., dissociate) into the sodium and chlorine atoms. The simplest possible chemical reaction, however, involves conversion of some molecule A into a different form B. To describe this reaction, we write the following equation:

$$A = B. \tag{3.1}$$

By convention, whatever molecules appear on the left hand side of the equation are called the "reactants" and the molecules appearing on the right are the "products." So, in our example, A is the reactant and B is the product. Since no other molecules are involved in this process, A and B must consist of the same atoms. That is, they must be described by the same chemical formula. The only difference between A and B can then be in the manner in which those atoms are arranged in the molecule. Figure 3.1 shows two examples: In the first one, we have the molecule of 1,2-dichloroethylene, which is described by the chemical formula $C_2H_2Cl_2$. This molecule can exist in two possible conformations (called isomers), which are our molecules A and B. The conversion between A and B in this case is what chemists refer to as isomerization. The second example involves a much more complicated process, in which a disordered molecular chain A undergoes a rearrangement, whereby it forms a well-defined structure, the protein molecule B. This process is commonly referred to as protein folding, where A represents the unfolded protein state and B is the folded one. Biological molecules, such as proteins and RNA, have to fold in order to function, and so the processes of this type are of great interest to biophysicists and biochemists.

At this point you may ask: Why are A and B the only two states used to describe this process? Clearly, there is a continuum of different shapes that a molecule can assume. For example, it seems likely that to go from one planar configuration in Fig. 3.1 to the other, the molecule $C_2H_2Cl_2$ must be twisted 180 degrees around the $C = C$ bond, thus requiring it to assume a series of nonplanar shapes along the way.

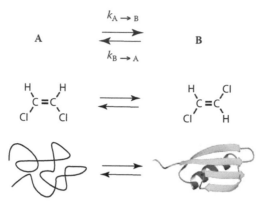

FIGURE 3.1 Examples of a reversible reaction A=B: isomerization of a small molecule and protein folding/unfolding. The second process involves self-structuring of a long molecular chain, typically consisting of hundreds or thousands of atoms, which ends up with a folded protein molecule. The atomistic details are not shown in the second case.

Perhaps, then, a scheme of the form $A = I_1 = I_2 = \cdots = B$, where I_n represents all those nonplanar intermediates (the number of which is, strictly speaking, infinite), would make more sense. Likewise, in our second example the unfolded protein state A does not even represent any unique molecular arrangement. Rather, A describes a collection of random-looking molecular shapes each resembling a long piece of spaghetti. Why do we lump all those into a single state? These are very interesting and important questions. The fundamental reasons why a collection of states can, in certain cases, be represented as a single state will be explained later on (Chapter 5). For now let us adopt a more pragmatic view that is based on how different molecular configurations are observed in practice. A chemist tells the difference between A and B by observing some physical property of the molecule. For example, A and B may absorb light at different wavelengths, i.e., they have different colors. If only two colors can be experimentally observed, we call the chemical containing the molecules of one color A and the other color B. If more colors are seen, then we need more states, or molecular species, to describe the process. It turns out that in many cases only two colors, rather than a continuum of colors, are observed. Whenever this is the case we use our reaction scheme (3.1).

There is another empirical observation that will be adopted here (and justified in subsequent chapters). It is quite common for the reactions described by Eq. 3.1 that if one mixes N_A molecules of A and N_B molecules of B, then these numbers will evolve in time according to the following differential equations

$$\frac{dN_A}{dt} = -k_{A \to B} N_A + k_{B \to A} N_B$$

$$\frac{dN_B}{dt} = -k_{B \to A} N_B + k_{A \to B} N_A. \qquad (3.2)$$

Chemists call the time evolution described by these equations "first order kinetics." These equations are not exact. Indeed, they cannot be exact since they describe

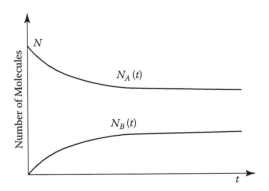

FIGURE 3.2 Starting with the pure reactant A, the amount of the reactant decreases and the amount of the product B increases until they reach their equilibrium values.

continuously changing functions while the number of molecules is an integer number. Nevertheless, as long as the total number of molecules in the system is very large, they are a reasonable approximation. The quantities $k_{A \to B}$ and $k_{B \to A}$ are called the *rate constants* or *rate coefficients* for going from A to B or B to A. They are not really constants in that they typically depend on the temperature and, possibly, on other properties of the system. Nevertheless, as silly as the phrase "temperature dependence of the rate constant" may sound, this is something you will get used to if you hang around with chemists long enough. What chemists really mean is that those quantities are independent of time. But even this is not necessarily true since it is conceivable that, in the above scheme, $k_{A \to B}$ and $k_{B \to A}$ could be time-dependent if, for example, experimental conditions are varied with time. To avoid any ambiguities I will use the term "rate coefficient" from now on.

Let us now examine what the solution of Eq. 3.2 looks like. Suppose that we start with molecules of A only and no molecules of B. For example, we could heat up our solution of proteins, which typically results in all of them being unfolded; then we could quickly reduce the temperature of the solution and watch some of them fold. That is, our initial condition is $N_A(0) = N$, $N_B(0) = 0$. The solution is then given by:

$$N_B(t) = N \frac{k_{A \to B}}{k_{A \to B} + k_{B \to A}} (1 - e^{-(k_{A \to B} + k_{B \to A})t})$$
$$N_A(t) = N - N_B(t) \tag{3.3}$$

and is schematically shown in Figure 3.2.

We see that, at first, the number of molecules of A decreases and the number of molecules B increases. Eventually, as $t \to \infty$, these amounts attain constant values, $N_A(\infty)$ and $N_B(\infty)$. This situation is referred to as *chemical equilibrium*. The equilibrium amounts of A and B satisfy the relationship

$$k_{A \to B} N_A(\infty) = k_{B \to A} N_B(\infty), \tag{3.4}$$

which, according to Eq. 3.2, ensures that the equilibrium amounts stay constant, $dN_A/dt = dN_B/dt = 0$.

Imagine now that we can observe each individual molecule in the mixture. According to our initial condition, at $t = 0$ they are all in state A. Eq. 3.3 then predicts that the population of molecules in A will start decreasing as a function of time, following a smooth curve such as the one shown in Fig. 3.2. How can we interpret this behavior in terms of the underlying behavior of each individual molecule? Perhaps the molecules, initially all in state A, are "unhappy" to be so far away from the equilibrium situation and so they are all inclined to jump into B, in an organized fashion, until sufficient number of molecules B are created. This, however, seems highly unlikely. Indeed, for this to happen the behavior of a molecule should be somehow influenced by the number of other molecules in the states A and B. Molecules interact via distant-dependent forces. It is, of course, not inconceivable that interaction between two neighboring molecules could make them undergo a transition to another state in a concerted fashion. However this can only happen when the molecules are close enough. If the experiment is performed in a large beaker, the behavior of two sufficiently distant molecules should be uncorrelated. Moreover, any cooperativity among molecules would disappear as the number of molecules per unit volume becomes smaller and smaller and, consequently, the average distance between the molecules increases. Since Eq. 3.2 is known to be satisfied even for very dilute solutions of molecules, the hypothesis that the molecules somehow conspire to follow these equations must be wrong.

On the contrary, Eq. 3.2 is consistent with the view that the molecules are entirely independent of one another and each is behaving in a random fashion! Indeed, this is illustrated in Figure 3.3, where we have $N = 6$ molecules (the trajectories of only 3 of them are shown). Each molecule jumps randomly between the states A and B. I have generated a trajectory for each on my computer using a simple random number generator (a computer program that spits out random numbers). Of course I had to know some additional information: For example, it is necessary to know something about the average frequency of the jumps. The precise mathematical description of the random jump process will be given below.

The trajectory of each molecule was generated independently. The molecules, however, were "synchronized" at the beginning of the calculation. That is, they are all in state A at $t = 0$. If we wait for some time, some of them will make a transition to B so that N_A will decrease (and N_B will increase). Indeed, such a decrease is observed in Fig. 3.3. Although the shape of the curve $N_A(t)$ is somewhat noisy, it is close to the curve obtained from Eq. 3.3. If I repeat this calculation (i.e., generate new random-jump trajectories for each molecule) I will get another $N_A(t)$ curve. Most of the time, it will be close to Eq. 3.3 although, occasionally, significant deviations may be observed. The noise will however go away if we repeat the same experiment with a much larger number of molecules. Even though each individual molecule jumps between its two states in a random fashion, a large collection (or, as physicists call it, an *ensemble*) of such molecules behaves in a predictable way. Such a predictable behavior is not unusual when large numbers of randomly behaving objects are involved. Your insurance company, for example, cannot predict each individual car accident occurring to their clients yet they can estimate the total number of accidents. They would not know how much to charge you for your policy otherwise. Likewise, although each molecule of air is equally likely to be anywhere in your house, you probably do not

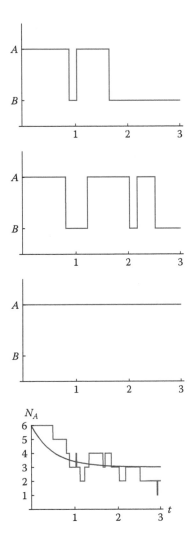

FIGURE 3.3 Six molecules were initially prepared in the reactant state (A). Each molecule proceeds to jump, in a random fashion, between A and B. Only the trajectories of three of them are shown. This results in a decaying population of state A, N_A, which agrees with the prediction of Eq. 3.3 shown as a smooth line.

need to worry about suffocating in your sleep as a result of all the molecules gathering in the kitchen.

Our picture of randomly behaving, independent molecules also explains the origin of chemical equilibrium in the system. At $t = 0$ all the molecules are in state A. Since there are no molecules in state B, the only transitions we will see early on are those from A to B. As the time goes by, there will be more molecules in B and, consequently, transitions from B to A will begin to occur. Once the number of A $\rightarrow$ B transitions is, on the average, balanced by that of B $\rightarrow$ A transitions, the total number of molecules in A or B will stay nearly constant. Equilibrium, therefore, results from a dynamic balance between molecules undergoing transitions in both directions.

Let us emphasize that equilibrium here is a property of the *ensemble* of the molecules, not of each individual molecule. That is, each molecule does not know whether or not it is in equilibrium. In other words, if we were to examine the sequence of random jumps exhibited by any specific member of the ensemble, it would not appear in any way special at $t = 0$ or at any value of time t. The nonequilibrium situation at $t = 0$ results from synchronization of all the molecules, i.e., forcing them all to be in state A. This creates a highly atypical (i.e., nonequilibrium) state of the entire ensemble for, while there is nothing atypical about one particular molecule being in state A, it is highly unlikely that they all are in A, unless forced to by the experimentalist. Starting from this initial state, each molecule proceeds to evolve randomly and independently of other molecules. At sufficiently short times t, there still is a good chance that many molecules have remained in the initial state A so their states are correlated, not because they interact with one another but simply because the ensemble has not forgotten its initial state. Once all of them have had enough time to make a few jumps, they are all out of sync (see Fig. 3.3), any memory of the initial state of the ensemble is lost, and equilibrium has been attained.

So far I have been vague about the specifics of the random-jump process exhibited by each molecule. To develop a mathematical description of this process, divide Eq. 3.2 by the total number of molecules $N = N_A + N_B$. As our molecules are independent, it then makes sense to interpret $w_{A,B} = N_{A,B}/N$ as the probabilities to be in states A or B. This gives

$$\frac{dw_A}{dt} = -k_{A \to B} w_A + k_{B \to A} w_B$$

$$\frac{dw_B}{dt} = -k_{B \to A} w_B + k_{A \to B} w_A. \qquad (3.5)$$

According to these equations, if the time is advanced by a small amount δt then the probability of being in A will become:

$$w_A(t + \delta t) = (1 - k_{A \to B}\delta t)w_A(t) + (k_{B \to A}\delta t)w_B(t). \qquad (3.6)$$

We can interpret this as follows:

$$w_A(t + \delta t) = \text{\small (conditional probability to stay in A having started from A)} \times \text{\small (probability to start in A)} +$$

$$+ \text{\small (conditional probability to make transition from B to A)} \times \text{\small (probability to start in B)}.$$

Therefore if the molecule is found in state A, the probability for it to make a jump to B during a short time interval δt is equal to $k_{A \to B}\delta t$. The probability that it will stay in A is $1 - k_{A \to B}\delta t$. Loosely speaking, the rate coefficient $k_{A \to B}$ can then be interpreted as the probability of making the transition from A to B per unit time.

By advancing the time in small steps δt and deciding whether or not to jump to the other state as described above, a single-molecule trajectory can be created, which corresponds to specified values of the rate coefficients $k_{A \to B}$ and $k_{B \to A}$. This way, you can generate your own version of Figure 3.3. Suppose, for example, that at $t = 0$

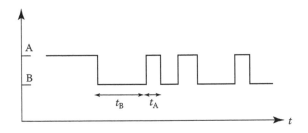

FIGURE 3.4 A typical trajectory of a molecule undergoing transitions between states A and B.

the molecule is in A. Advance time by δt. Now choose the new state to be B with the probability $k_{A \to B}\delta t$ and A with the probability $1 - k_{A \to B}\delta t$. And so on. Whenever the molecule is in state B, upon advancement of time by δt the new state becomes A with the probability $k_{B \to A}\delta t$ and remains B with the probability $1 - k_{B \to A}\delta t$. This procedure creates a discrete version of the trajectory shown in Fig. 3.4, with the state of a molecule specified at $t = 0, \delta t, 2\delta t, \ldots$. It will become continuous in the limit $\delta t \to 0$.

A traditional chemical kineticist deals with curves describing the amounts of various molecules such as the ones shown in Fig. 3.2. Given the experimentally measured time evolution of $N_A(t)$ and $N_B(t)$, he or she can attempt to fit these dependences with Equations 3.2 or 3.3. If such a fit is successful, it will provide an estimate for the rate coefficients $k_{A \to B}$ and $k_{B \to A}$. In contrast, a single-molecule chemist is faced with a dependence like the one shown in Fig. 3.4. Because this dependence is a random process, it is not immediately obvious how to extract the rate coefficients from such experimental data. Fortunately, Fig. 3.4 happens to contain the same information as Fig. 3.2. To find out how to infer this information, consider the time t_A (or t_B) the molecule dwells in A(or B) before it jumps to the other state (Fig. 3.4). Let

$$w_{A \to B}(t_A)dt_A$$

be the probability that the molecule will make a transition from A to B between t_A and $t_A + dt_A$, where the clock measuring t_A was started the moment the molecule has made a transition from B to A. To find this quantity experimentally, we could follow the molecule's trajectory (Fig.3.4) for a while. Whenever it jumps to state A we start our clock and whenever it jumps back to B we stop the clock and record the dwell time t_A. Suppose we repeat this experiment M times, where $M \gg 1$. A histogram of the dwell times t_A provides an experimental estimate of $w_{A \to B}(t_A)$. The same quantity could be measured in a different manner: Instead of making M measurements on one molecule, we pick M identical molecules, which all happened to be in state A at $t = 0$.[1] We follow the state of each molecule until it jumps to the state B, and record the time t_A it took for the jump to occur. Since we are only interested in the time t_A

---

[1] The astute reader may object that the two experiments are not equivalent unless we demand that each of the M molecules has just made a transition from B to A at $t = 0$. It is, however, sufficient to demand that each molecule is found, at $t = 0$, in A, regardless of when it has arrived in this state. This is because, in the process we are describing here, the fate of the molecule depends only on the *current state* and not on the molecule's past.

and not in what happens to each molecule afterwards, we can discard the molecule as soon as it jumps, for the first time, to B. The number of molecules M_A still remaining in state A obeys the equation

$$dM_A/dt_A = -k_{A \to B} M_A,$$

with the initial condition $M_A(0) = M$. This equation describes an irreversible first-order reaction

$$A \to B$$

where each molecule disappears (i.e., gets discarded) as soon as it arrives in B. The solution of this equation is

$$M_A(t_A) = M \exp(-k_{A \to B} t_A). \tag{3.7}$$

This quantity decreases with time t_A because more and more molecules jump to B. By the time $t_A + dt_A$, for example, M_A has decreased by the amount equal to the number of molecules that have left the state A during the time interval dt_A:

$$M_A(t_A) - M_A(t_A + dt_A) = -dM_A = k_{A \to B} M_A dt_A = M k_{A \to B} \exp(-k_{A \to B} t_A) dt_A.$$

To obtain the probability that the transition happens between t_A and $t_A + dt_A$, we simply divide this result by M:

$$w_{A \to B}(t_A) dt_A = k_{A \to B} \exp(-k_{A \to B} t_A) dt_A. \tag{3.8}$$

Notice that, according Eq. 3.7, the exponential function $\exp(-k_{A \to B} t_A)$ is equal to the fraction of molecules that remain in state A until the time t_A, or, equivalently, to the probability to survive in state A. Eq. 3.8 can then be interpreted as the transition probability $k_{A \to B} dt_A$, conditional upon finding the molecule in A at t_A, times the probability to survive in state A until t_A. As expected, this transition probability satisfies the condition $\int_0^\infty w_A(t_{A \to B}) dt_A = 1$, which implies that, if one waits long enough, the molecule will undergo a transition to B, with certainty. "Long enough" means much longer than the average dwell time in A, which is given by

$$\langle t_A \rangle = \int_0^\infty t_A w_{A \to B}(t_A) dt_A = 1/k_{A \to B}. \tag{3.9}$$

The latter relationship shows that the rate coefficients are directly related to the mean dwell times spent in each state and so they can be immediately inferred from a long trajectory such as the one shown in Figure 3.4. These dwell times further satisfy the relationship

$$\langle t_B \rangle / \langle t_A \rangle = k_{A \to B} / k_{B \to A}. \tag{3.10}$$

Compare this with Eq. 3.4, which predicts:

$$\langle N_B(\infty) \rangle / \langle N_A(\infty) \rangle = k_{A \to B} / k_{B \to A}. \tag{3.11}$$

The similarity between the two ratios is not surprising: It is intuitively appealing that the equilibrium number of molecules found in state A should be proportional to the time each given molecule spends in this state. More generally, this is the *ergodicity principle* at work. This principle declares that any time average performed over a long trajectory of a molecule should be identical to an average over an equilibrium ensemble of such molecules. Consider the equilibrium probability of being in state A, $w_A(\infty)$. It can be calculated as the ensemble average, i.e., the fraction of molecules found in A:

$$w_A(\infty) = \frac{N_A(\infty)}{N_A(\infty) + N_B(\infty)}.$$

It can also be calculated as the time average, i.e., the fraction of the time the molecule spends in A:

$$w_A(\infty) = \frac{\langle t_A \rangle}{\langle t_B \rangle + \langle t_B \rangle}.$$

In view of Eqs. 3.10 and 3.11, the two methods give the same result:

$$w_A(\infty) = \frac{k_{B \to A}}{k_{A \to B} + k_{B \to A}}.$$

Although these findings are plausible, we note that there are many systems that do not behave in an ergodic way. The ergodicity principle will be further discussed in the next chapter.

Exercise

Consider a molecule that can exist in three forms, A, B, and C. The interconversion between them is described by the scheme

$$A \underset{k_{B \to A}}{\overset{k_{A \to B}}{\rightleftharpoons}} B \underset{k_{C \to B}}{\overset{k_{B \to C}}{\rightleftharpoons}} C.$$

According to this scheme, for example, the numbers of A and B molecules obey the equations

$$dN_A/dt = -k_{A \to B} N_A + k_{B \to A} N_B$$

and

$$dN_B/dt = k_{A \to B} N_A + k_{C \to B} N_C - k_{B \to A} N_B - k_{B \to C} N_B.$$

What are the average times $\langle t_A \rangle$, $\langle t_B \rangle$, and $\langle t_C \rangle$ the molecule dwells, respectively, in its A, B, and C forms? Compare the ratio $\langle t_B \rangle / \langle t_A \rangle$ to the ratio $N_B(\infty)/N_A(\infty)$ of the equilibrium populations. Are they the same? If not, what causes the discrepancy?

Our example shows that a single-molecule measurement can provide the same information (i.e., the rate coefficients) as the traditional, bulk chemical kinetics. But if both give the same information, why do single-molecule measurements? As discussed in the preceding chapter, single-molecule experiments require rather involved microscopy and single-photon detection and so they can be difficult and expensive. Moreover, real single-molecule data rarely look like our idealized Figure 3.4. Rather, real single-molecule trajectories are plagued by noise and are often too short to allow reliable statistical analysis required to accurately estimate the underlying kinetic parameters such as the rate coefficients.

The above are just a few reasons for the skepticism encountered by the single-molecule experimenters in the early stages of the field. There are, however, many advantages to measuring chemical kinetics the single-molecule way. Those advantages will become more clear later on in this book. The simple two-state example considered here, however, already reveals one useful property of single-molecule measurements. Suppose the transition from A to B is much faster than from B to A, i.e., $k_{A \to B} \gg k_{B \to A}$. Then, according to Eq. 3.11, we have $N_A \ll N_B$ in equilibrium. What if we are specifically interested in the properties of the molecule A? As the equilibrium mixture of A and B mostly consists of B, whatever experimental measurement we choose to make will report mostly on the properties of B rather than A. If we want to specifically study A, first of all we will need to find a way to get rid of B by driving all (or most) of our molecules to the state A. Even if we have succeeded in doing so, we only have a very short time interval, $\Delta t \ll 1/k_{A \to B}$, before most of the molecules will jump back to B. A single-molecule approach resolves this problem: As we observe a molecule's trajectory, we know what state it is in. If we are specifically interested in A, then we simply interrogate the molecule while it is in A.

A related advantage of single-molecule experiments is that they do not require synchronization of a large number of molecules to study their time-dependent properties. Indeed, in order to measure $k_{A \to B}$ and $k_{B \to A}$ using a bulk technique, it is usually necessary that one starts away from equilibrium. For example, to observe the curve of Fig. 3.3 it is necessary to initially synchronize the molecules by forcing all of them into the state A. As mentioned earlier, this sometimes can be achieved by quickly changing the conditions of the experiment. For example, one can quickly change the temperature or the pressure in the system. In doing so, it is necessary to ensure that the preparation process itself is much faster than the process one is trying to observe (e.g., the conversion of A to B and back to A). If, for instance, the rate coefficients $k_{A \to B}$ and $k_{B \to A}$ are of order of $10^3 s^{-1}$, then one will have to devise a way of heating the system up or cooling it down in less than a millisecond. This is not easy, especially if A and B are mixed in a beaker of a macroscopic size! Single-molecule experiments sidestep this difficulty altogether by observing the trajectories of molecules under equilibrium conditions, one molecule at a time.

4 How Molecules Explore Their Energy Landscapes

What is really strange and new in the Brownian movement is, precisely, that it never stops. At first that seems in contradiction to our every-day experience of friction. If for example, we pour a bucket of water into a tub, it seems natural that, after a short time, the motion possessed by the liquid mass disappears... What we observe ... is not a cessation of the movements, but that they become more and more chaotic, that they distribute themselves in a fashion the more irregular the smaller the parts.

Jean Perrin, *Brownian motion and molecular reality*

4.1 THE POTENTIAL ENERGY SURFACE

A diatomic molecule, such as CO, N_2, or O_2, consists of two nuclei and a handful of electrons. The simplest example is the hydrogen molecule, H_2, consisting of two protons and two electrons. Electrostatic interactions involving the charged nuclei and the electrons lead to an effective attraction between the two nuclei, which is the very reason the molecule is formed. This attraction can often be described using the following picture: Imagine fixing the distance R between the two nuclei. Now consider the molecule's electrons interacting with each other while also subjected to the electric field created by the nuclei. The behavior of those electrons is described by quantum mechanics. Specifically, quantum mechanics predicts in this case that, for each value of R, the molecule can take on discrete energy levels $E_0(R)$, $E_1(R)$, etc. If the temperature of the system is not too high then the higher energy levels are often thermally inaccessible.[1] We thus concentrate on the lowest energy level $E_0(R)$. By adding this energy to the energy of internuclear electrostatic interaction, we obtain a function $V(R)$, which can be thought of as an effective potential felt by the nuclei. The typical $V(R)$ looks like the curve shown in Figure 4.1. In the limit $R \to \infty$ the two atoms do not interact with one another so that $V(R)$ approaches a constant value. R being too small results in a repulsive interaction such that $V(R)$ becomes very high. The existence of a minimum of $V(R)$ at an intermediate value $R = R_0$ signifies a chemical bond. That is, mutual approach of two atoms lowers their total potential energy. Conversely, separating the two atoms requires expending energy.

Finding the precise shape of $V(R)$ involves solving a quantum mechanical many-body problem involving all the electrons of the molecule. We will not be concerned with such a calculation here but rather assume that such a function is "given" to us.

---

[1] According to the laws of statistical mechanics (see Appendix B) the probability of having an energy equal to E_1 at temperature T is proportional to $\exp[-E_1/(k_BT)]$, where k_B is Boltzmann's constant. This probability is negligible as compared to the probability of having an energy equal to E_0 provided that we have $E_1 - E_0 \gg k_BT$.

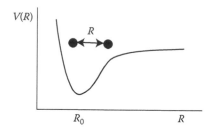

FIGURE 4.1 Potential energy of a diatomic molecule as a function of the interatomic distance.

In practice, precise calculation of potential energies often involves high computational cost and so empirical models are commonly used instead. The simplest empirical model replaces $V(R)$ by its quadratic expansion near the energy minimum R_0, i.e.,

$$V(R) \approx V(R_0) + (1/2)V''(R_0)(R - R_0)^2.$$

This model effectively replaces the chemical bond between the atoms by a Hookean spring. Despite its simplicity, this model can be remarkably useful. For example, it predicts the specific heat of gases consisting of diatomic molecules with a remarkably high accuracy. It however cannot predict the breaking of a chemical bond, as the two atoms connected by a truly Hookean spring will never escape each other's attraction.

For a molecule, or a collection of several molecules comprised of three or more atoms, we can write the potential energy as a function of the positions of each atom (or, more precisely, the positions of the nuclei),

$$V = V(x_1, y_1, z_1, x_2, y_2, z_2, \ldots).$$

We call this function the "potential energy surface" (PES) of the molecule. If the number of relevant degrees of freedom is equal to one then this surface amounts to a curve like the one shown in Fig. 4.1. For two degrees of freedom such a surface can be visualized as a topographic map shown in Fig. 4.2. It is much harder to visualize a multidimensional potential energy surface.

4.2 WHAT ARE THE EQUATIONS OF MOTION OBEYED BY A MOLECULE?

To proceed further, we need to specify the equations governing the motion of the molecule(s). The simplest possibility is to assume that molecules obey the laws of classical mechanics. That is, the trajectory of each atom is described by Newton's second law, which can be written as a differential equation of the form:

$$m_i \ddot{x}_i = -\partial V(x_1, y_1, z_1, \ldots)/\partial x_i, \tag{4.1}$$

where m_i is the mass of the i-th atom. The left hand side of Eq. 4.1 is the atom mass times its acceleration component along x while the right hand side is the force acting on this atom and projected on the x-axis. It should be remembered that writing Eq. 4.1

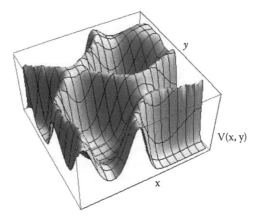

FIGURE 4.2 Potential energy, as a function of two degrees of freedom, forms a surface.

involves multiple approximations. First of all, the potential V was defined above as the molecule's electronic energy in a situation where its nuclei are clamped in a fixed configuration and the electrons are in their ground state. Moving nuclei, however, subject the electrons to a time-dependent perturbation that may cause the electrons to make transitions to different states. Fortunately, the above model is often (but not always!) reasonable because the nuclei move much slower than the electrons do; thus, to a good approximation, the electrons effectively see static nuclei at any given time.

Second, Eq. 4.1 completely ignores quantum mechanical effects with regard to the nuclear motion. Such neglect of quantum mechanics is, of course, not uncommon in many fields of science and engineering: After all, quantum mechanics is hardly needed to describe the motion of a plane or a car. Using Newtonian mechanics instead is justified because planes and cars are massive bodies, whose de Broglie wavelengths[2] are much shorter than any practically relevant length scale. While considerably heavier than electrons, atomic nuclei, however, are not nearly as massive as macroscopic objects that surround us. Consequently, they cannot always be viewed as classical particles. Quantum mechanical effects are especially significant for lighter atoms, such as hydrogen and helium. For example, classical mechanics will never explain why, unlike other materials, helium never crystallizes into a solid (at atmospheric pressure), even at zero temperature.

These caveats notwithstanding, let us proceed, cavalierly, with exploring the consequences of Eq. 4.1 on the dynamics of a molecule. One immediate consequence is that the total energy of the molecule,

$$E = \frac{p_1^2}{2m_1} + \frac{p_2^2}{2m_2} + \cdots + V(x_1, y_1, z_1, x_2, y_2, z_2, \ldots),$$

---

[2] The de Broglie length of an object can be thought of as a lengthscale at which quantum mechanical effects become important. It is inversely proportional to the object's momentum and, therefore, its mass. As a result, bodies with large masses are well described by Newtonian mechanics. For microscopic particles, quantum effects become more pronounced at low temperatures, where the typical thermal velocity and, therefore, momentum become small.

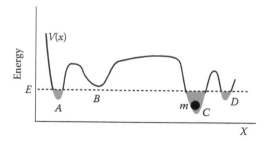

FIGURE 4.3 At a fixed energy E, the configuration space available to a particle moving in a potential $V(x)$ is disjointed. For example, a classical particle of energy E found initially in the well C will never get to the wells A or D.

is conserved. Here p_i is the total momentum of the i-th nuclei. To simplify the discussion, let us pretend that the molecule has only one degree of freedom, x, and is subjected to a one-dimensional potential $V(x)$ like the one shown in Figure 4.3. Then we simply have

$$E = \frac{p^2}{2m} + V(x).$$

When E is fixed, only the regions where $V(x) < E$ are accessible to the system. This is illustrated in Fig. 4.3, where the system can occupy three disjointed regions (A,C, and D). Since these regions are separated by potential barriers, for which $V(x) > E$, the molecule then remains trapped forever[3] in whichever region it has started. It therefore fails to sample all of its energetically accessible conformational space.[4] In such a situation, it is said that the molecule lacks *ergodicity* or that its motion is nonergodic. A more formal definition of ergodicity is that, for an ergodic system, the time average of any physical quantity is equal to the ensemble average. For example, we could measure the position of the system, $x(t)$, as a function of time and define the average position as a time average over a sufficiently long time interval τ,

$$\langle x \rangle = \tau^{-1} \int_0^\tau dt\, x(t).$$

It is, however, clear that, no matter how long the time τ is, the value of this average would depend on the initial condition and would be different depending on which region of space (A,C, or D) the system has started from at $t = 0$. On the other hand, we could also define an ensemble average, i.e., the average of x over all possible energetically accessible states of the system. Unlike the time average, the ensemble average is unique and includes all physically possible values of x.

The outcome of a measurement performed on a nonergodic system is not reproducible, unless the initial condition is precisely known. Therefore, in contrast to the

---

[3] Note that quantum mechanics permits the system to penetrate classically forbidden barriers via the quantum tunneling effect. Thus "forever" should be replaced by "for some time." This time is dependent on the height and the width of the barrier.

[4] We refer to all possible vectors $(x_1, y_1, z_1, x_2, y_2, z_2, \ldots)$ describing possible conformations of a molecule as its conformational space. Here, "all conformational space" is a fancy way of saying "all values of x."

ergodic scenario discussed in Chapter 3, single-molecule measurements may not provide the same information as ensemble measurements. This is actually good news, since the purpose of single-molecule measurements is not to reproduce ensemble measurements but to learn something new. For example, a single trajectory observed for the system depicted in Figure 4.3 will tell us about the properties of an individual region, A, C, or D, rather than the system's properties averaged over all the three regions.

4.3 STOCHASTICITY IN THE DYNAMICS OF INDIVIDUAL MOLECULES

Equation 4.1 assumes that the molecule of interest is isolated from its surroundings. Such an assumption is almost never realistic as any molecule we choose to study will inevitably collide with the surrounding molecules and/or with the walls of its container. To account for the molecule's interaction with its surroundings, one may consider the dynamics of an extended system that consists of the molecule of interest plus the molecules inside the container plus the molecules of the container itself.[5] The dynamics of such a super-system is still described by Newton's equations of motion of the form of Eq. 4.1. However, the potential V is now a function of the coordinates of all the molecules inside the container as well as the molecules forming the container walls. As a result, we no longer can predict the behavior of the molecule of interest without solving an astronomically large system of differential equations describing every other molecule of the super-system. A useful approximation is to replace such an astronomic number by a still very large yet manageable (say, one thousand or one million) number of molecules. This keeps the problem computationally tractable yet, one hopes, provides a better approximation of the reality than the model comprised of an isolated molecule. This kind of approach is adopted by the popular class of computer simulation methods called *molecular dynamics*, which solve Newton's equations of motion for relatively large assemblies of atoms.

It seems plausible that having many particles in a system would make the dynamics of its molecules more ergodic, as compared to the isolated molecule case discussed above. Indeed, while one molecule may not have enough energy to overcome an energy barrier and go from one conformation to another, as in Fig. 4.3, it may gain such energy as a result of a collision with another molecule. Beyond such plausible arguments and with the exception of a few special cases, ergodicity of a system is usually very difficult to prove or disprove. For this reason, the existence of ergodicity in a system is usually stated as an *ergodicity hypothesis*. Ergodicity in a strict mathematical sense should also be distinguished from its more practical view: A block of wood, for instance, is, for any practical purpose, a patently non-ergodic system in that its constituent cellulose molecules fail to explore vast regions of accessible conformational space. Specifically, unless the temperature is raised to a point that the wood starts burning,

---

[5] As the container interacts with its own surroundings, one should not, strictly speaking, stop with the container but rather proceed including more and more molecules until the entire Universe is taken into consideration.

they fail to undergo the (thermodynamically favorable) oxidation process and turn into carbon dioxide and water. The apparent non-ergodicity in this case has to do with the fact that the oxidation reaction is too slow, which, in turn, is a consequence of a very high energy barrier intervening between the reactants and the products of this reaction (see Chapter 5). It is possible to speed up this reaction by throwing the wood into a fire. Alternatively, one would have to wait a very long time before the behavior would appear ergodic. We thus conclude that the ergodic behavior, from a practical viewpoint, is often a matter of timescale. Many processes will appear non-ergodic simply because the observation timescale is too short to allow full exploration of the conformational space.

Solving Newton's equations of motion for a large assembly of atoms only to describe the behavior of its small subsystem is expensive. A less costly alternative is to resort to a phenomenological model that accounts for the interaction of the system of interest with its surroundings in a simplified way. I will now describe one such model. For simplicity, I will again assume that the configuration of the system of interest can be described in terms of a single degree of freedom x. To be specific, let us pretend that the system of interest is a molecule in solution. In the absence of the solvent, the molecule's potential energy is $V(x)$ and its motion is governed by Newton's second law, i.e.,

$$m\ddot{x} = -V'(x).$$

How will the solvent molecules affect this motion? First of all, it may effectively change the interaction described by $V(x)$. For example, molecules that are stable in air may dissociate into fragments when placed in solution; the solvent then must effectively alter the attractive forces within those molecules. We thus introduce an *effective* potential $\tilde{V}(x)$ describing the environment-mediated interactions within the molecule. The physical significance of this effective potential will become more clear in Chapter 8. Second, interactions with the solvent lead to dissipative, frictional forces. For example, when a macroscopic body moves in a liquid, it experiences a viscous drag force acting in the direction opposite the direction of its movement. The magnitude of this force is proportional to the velocity (Stokes law), where the proportionality coefficient depends on the geometry of the body and on the properties of the liquid (specifically, its viscosity). Although this law can be rigorously justified only for objects that are large enough that the molecular structure of the liquid is unimportant, it is often empirically found to apply even to microscopic entities such as molecules.[6] Therefore, we write

$$m\ddot{x} = -\tilde{V}'(x) - \eta\dot{x},$$

where η is a coefficient that quantifies friction.

This equation, however, is not realistic as a description of molecular motion because it predicts that any molecule will eventually stop moving. Indeed, by multiplying both sides of this equation by $\dot{x}$, it can be rewritten as $d[m\dot{x}^2/2 + V(x)]/dt = -\eta\dot{x}^2$, which

---

[6] The viscous drag force can be viewed as the lowest-order term in the Taylor expansion of the resistance force as a function of velocity. It is dominant provided the velocity is low enough that higher-order terms can be neglected.

shows that the total energy of the molecule will always decrease until the molecule comes to rest (where $\dot{x} = 0$). What is missing in our picture is the thermal motion to keep our molecule agitated. In a crowded environment such as a liquid, the molecule is subjected to incessant kicks exerted by its neighbors. As those kicks are produced by many different molecules, they appear to be random and so we can model them as a random time-dependent force $R(t)$, whose mean value is zero, $\langle R \rangle = 0$.[7] We have arrived at what is called the Langevin equation:

$$m\ddot{x} = -\tilde{V}'(x) - \eta\dot{x} + R(t). \tag{4.2}$$

In a liquid, $R(t)$ is a result of interaction with many other molecules. Therefore it is the sum of many random numbers. If those numbers are independent, then the central limit theorem (see Appendix A) asserts that the values of R have a Gaussian distribution.

Consider now the temporal properties of the function $R(t)$. If we tabulate the values of $R(t)$ every δt picoseconds, we will end up with a succession of random numbers, $R_1 = R(\delta t)$, $R_2 = R(2\delta t)$, $R_3 = R(3\delta t)$, ... at times δt, $2\delta t$, and so on. The simplest model for R would be to assume that its value at any given time has no memory of its earlier values. That is, R_i and R_j are statistically independent unless $i = j$. As shown in Appendix A, the mean of the product of two statistically independent quantities is equal to the product of their means. Therefore, for $i \neq j$, we have

$$\langle R_i R_j \rangle = \langle R_i \rangle \langle R_j \rangle = 0.$$

That is,

$$\langle R_i R_j \rangle = \langle R^2 \rangle \delta_{ij} \tag{4.3}$$

where δ_{ij} is Kronecker's delta-symbol defined by $\delta_{ij} = 1$ if $i = j$ and 0 if $i \neq j$.

The continuous-time process $R(t)$ is recovered when the time δt between successive kicks goes to zero. This results in a continuous equivalent of Eq. 4.3:

$$\langle R(t)R(t') \rangle = A\delta(t - t'). \tag{4.4}$$

Here $\delta(t)$ is a function that is nonzero only if $t = 0$ and A is some quantity that characterizes the strength of the random kicks. If we were to plot $\delta(t)$, it would appear as an infinitely narrow spike at $t = 0$. It is, perhaps, a bit difficult to fathom such a mathematical object, but we do not have to worry too much about it because, if we were to actually model, or even display, such a function on a computer, it would have a finite (though small) width. We thus imagine a physical approximation $\delta_\epsilon(t)$ to $\delta(t)$, a function that is sharply peaked around $t = 0$, such that the width of the peak is finite and equal to ϵ. In other words, $\delta_\epsilon(t) = 0$ if $|t| > \epsilon$. Now imagine making ϵ smaller and smaller. It turns out that, in order to keep the coefficient A finite, the height of the peak should increase at the same time, such that the total area under

---

[7] It is always possible to define the random force in such a way that its mean value is zero because, if this is not the case, the deterministic force $\langle R \rangle$ can be incorporated in the potential $\tilde{V}(x)$. In fact, the mean force exerted by the solvent is precisely what causes the difference between the vacuum potential $V(x)$ and $\tilde{V}(x)$. For this reason, $\tilde{V}(x)$ is often called "the potential of mean force."

the curve $\delta_\epsilon(t)$ stays constant. Taking the $\epsilon \to 0$ limit results in the famous function known as Dirac's delta-function. This function is equal to zero for $t \neq 0$, is infinite at $t = 0$, and satisfies the following condition

$$\int_{-\infty}^{\infty} \delta(t)dt = 1.$$

More generally, Dirac's function satisfies the relationship:

$$\int_{-\infty}^{\infty} f(t)\delta(t-a)dt = f(a) \tag{4.5}$$

for any function $f(t)$ and any real number a. It will be shown below that the above definition of the delta-function, indeed, results in a finite value of A.

Eq. 4.4 is an example of what is called an autocorrelation function. It is simply the mean value of the product of values of the function $R(t)$ taken at different moments of time. For an ergodic system—and Langevin dynamics happens to be ergodic, a statement that we will accept without proof—we do not need to specify whether the mean is calculated via ensemble or time averaging. In particular, assuming that $R(t)$ is a stationary random process, whose statistical properties do not change over time, its autocorrelation function depends only on the time difference, $\langle R(t)R(t') \rangle = \langle R(0)R(t'-t) \rangle$, and can be computed as a time average over a single molecular trajectory observed over a sufficiently long time interval τ,

$$\langle R(0)R(t) \rangle = \tau^{-1} \int_0^\tau dt' R(t')R(t+t'). \tag{4.6}$$

So far the value of the kicking strength parameter A was not connected to any physical properties of the system. It must, however, have something to do with the temperature in the system as increasing its value will clearly enhance random motion of the molecule. Moreover, the random force R has essentially the same microscopic origin as the frictional force on the system, since both result from the interactions of the system with the solvent molecules. The two kinds of forces must then be related to one another. Indeed, if we remove any such interactions with the solvent by placing our molecule in vacuum, both the friction coefficient η and the strength of the noise A must vanish simultaneously. The relationship between these two parameters can be established if we demand that the statistical properties of the molecules as predicted by the Langevin equation must be consistent with equilibrium statistical-mechanics. In particular, the velocity $u = dx/dt$ of the particle obeying this equation must be distributed according to the Maxwell-Boltzmann distribution (see Appendix B) and so the mean kinetic energy of the molecule must be given by

$$\langle mu^2/2 \rangle = k_B T/2. \tag{4.7}$$

This should be true regardless of the shape of the potential $\tilde{V}$, including the case $\tilde{V}(x) = 0$ where there is no potential at all. Using Eq. 4.2 for this case, the velocity is seen to satisfy the first-order differential equation

$$m\frac{du}{dt} = -\eta u + R(t). \tag{4.8}$$

In the absence of the random force R, the solution of this equation would be given by

$$u(t) = u(0)e^{-\eta t/m}.$$

The solution to Eq. 4.8 can then be found by using the following ansatz:

$$u(t) = \tilde{u}(t)e^{-\eta t/m},$$

where $\tilde{u}(t)$ is some yet unknown function. Substituting this into Eq. 4.8, we find

$$\frac{d\tilde{u}}{dt}e^{-\eta t/m} = R(t)/m,$$

which gives

$$\tilde{u}(t) - \tilde{u}(0) = m^{-1}\int_0^t e^{\eta t'/m}R(t')dt',$$

and, finally,

$$u(t) = u(0)e^{-\eta t/m} + m^{-1}\int_0^t e^{-\eta(t-t')/m}R(t')dt'. \tag{4.9}$$

The first term in this expression decays exponentially, with a characteristic time

$$\tau_u = (\eta/m)^{-1}. \tag{4.10}$$

That is, the velocity "forgets" its initial value if $t \gg \tau_u$. Assuming this long-time limit, we now neglect the first term in Eq. 4.9, square $u(t)$, and take the average of the result:

$$\langle u^2 \rangle = m^{-2}\int_0^t dt'\int_0^t dt'' \exp[-\eta(2t - t' - t'')/m]\langle R(t')R(t'')\rangle.$$

Using Eqs. 4.4 and 4.5 and, again, assuming $t \gg \tau_u$, the above integral is readily evaluated to give

$$\langle u^2 \rangle = \frac{A}{2m\eta}.$$

Comparing this with Eq. 4.7, we conclude that

$$A = 2\eta k_B T. \tag{4.11}$$

This result, which is an example of a *fluctuation-dissipation relationship*,[8] confirms our expectation that the strength of the noise A should be related to the friction coefficient. It further shows that A is proportional to the temperature. Raising the temperature increases the thermal motion, leading to random forces of a higher magnitude. It can be further shown that, as long as the random noise satisfies Eq. 4.11, not

---

[8] Generally speaking, fluctuation-dissipation relationships establish the connection between equilibrium fluctuations in the system (such as the fluctuating force R) and the system's response to an external perturbation (here the friction coefficient η quantifies the mean velocity of the system, $\langle \dot{x} \rangle$, in response to a constant driving force F through the equation $\langle \dot{x} \rangle = F/\eta$).

only does the mean kinetic energy of the system satisfy Eq. 4.7 but the probability distribution of the coordinate x is precisely the Boltzmann distribution (see Appendix B) in the potential $\tilde{V}(x)$. That is, if we follow a very long trajectory $x(t)$, we will find that the fraction of time spent between x and $x + dx$ will be proportional to

$$w(x)dx = q^{-1} e^{\frac{-\tilde{V}(x)}{k_B T}} dx,$$

where q is an appropriate normalization factor. Likewise, the velocity of the particle obeys the Maxwell-Boltzmann distribution. This, in particular, implies that the dynamics predicted by the Langevin equation is ergodic.

4.4   PROPERTIES OF STOCHASTIC TRAJECTORIES

Now that we have a reasonable model for the dynamics of an individual molecule, we are ready to explore some of the properties of single-molecule trajectories. To this end, discussed below are several common scenarios.

4.4.1   FREE DIFFUSION

This is the case where the potential is flat, i.e., $\tilde{V}'(x) = 0$. The velocity u then satisfies the equation

$$\dot{u} = -\tau_u^{-1} u + R(t)/m, \qquad (4.12)$$

where τ_u is the velocity memory time defined by Eq. 4.10. An example of a computer-generated solution of this equation is shown in Fig. 4.4. We see that $u(t)$ fluctuates around zero. These fluctuations originate from the competition between the two components of the force acting on the system: In the absence of the random force, the velocity would relax exponentially to zero (cf. Eq. 4.9) while the random force kicks the system away from the equilibrium value $u = 0$.

To gain further insight into the properties of $u(t)$, it is helpful to consider its autocorrelation function

$$C_u(t) = \langle u(0)u(t) \rangle. \qquad (4.13)$$

Similarly to the autocorrelation of the force $R(t)$ (cf. Eq. 4.6), $C_u(t)$ is simply the mean value of the product of the velocities measured at different moments of time. This quantity can be used to inform us about the timescale over which this function remembers its previous values. Indeed, while, at $t = 0$,

$$\langle u(0)u(t) \rangle = \langle u^2 \rangle \equiv u_0^2$$

is finite, for times t long enough that $u(0)$ and $u(t)$ become statistically independent, the average of the product becomes equal to the product of the averages, $\langle u(0)u(t) \rangle = \langle u(0) \rangle \langle u(t) \rangle = 0$, and the autocorrelation function vanishes. To calculate this correlation function at an intermediate time t, we multiply Eq. 4.9 by $u(0)$ and take the average of the result. This gives

$$\langle u(0)u(t) \rangle = \langle u^2(0) \rangle e^{-\eta t/m} + m^{-1} \int_0^t \exp[-\eta(t-t')/m] \langle u(0)R(t') \rangle dt' = \frac{k_B T}{m} e^{-t/\tau_u}.$$

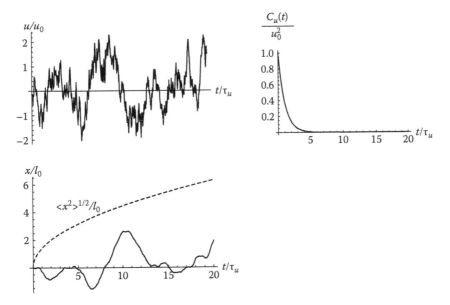

FIGURE 4.4 A typical time dependence of the velocity (upper left) and position (lower left) of a particle undergoing free diffusion. The root mean square distance traveled, as a function of time, is also shown in the lower left as a dashed line. The velocity autocorrelation function is plotted, as a function of time, in the upper right. The velocity is normalized by the root mean square thermal velocity $u_0 = \sqrt{k_B T/m}$ while the unit of distance l_0 is defined as $D\tau_u$, where the diffusion coefficient D is defined by Eq. 4.17.

The term containing the random force $R(t)$ has disappeared upon averaging because this random force is statistically independent of the initial velocity and because the average value of R is zero. Examining the result, we see that the time τ_u defined above also describes the timescale over which the memory of the previous values of $u(t)$ is lost.

Let us now turn to the time dependence of the system's coordinate $x(t)$. As we are presently considering motion in the absence of a spatially dependent potential, we can assume $x(0) = 0$ without any loss of generality. Since the velocity of the particle (and, hence, its sign) is remembered for a time that is of order τ_u, we can try replacing the continuous trajectory $x(t)$ by the following crude model: We envision a discrete random walk, in which the particle makes a step of length $\Delta x_i = u_i \tau_u$ either left or right every time interval τ_u. Here u_i is the velocity during the i-th step, which we assume to be approximately constant for the duration of the step. The total displacement of the molecule during time t is given by

$$x(t) = \Delta x_1 + \Delta x_2 + \cdots \Delta x_n = \tau_u(u_1 + u_2 + \cdots + u_n), \qquad (4.14)$$

where

$$n = t/\tau_u$$

is the number of steps. The mean displacement is obtained by averaging $x(t)$. This gives $\langle x(t) \rangle = 0$ since the velocity is equally likely to be positive or negative. A better measure of the distance traveled away from the origin would be the mean square displacement, $\langle x^2(t)\rangle$. Using Eq. 4.14,

$$\langle x^2(t)\rangle = \tau_u^2(\langle u_1^2\rangle + \langle u_2^2\rangle + \cdots + \cdots \langle u_n^2\rangle + \langle u_1 u_2\rangle + \cdots). \qquad (4.15)$$

Since the velocity memory is presumed lost after τ_u, the velocities during different steps are statistically independent and so the terms $\langle u_i u_j\rangle$ are zero unless $i = j$. We thus find:

$$\langle x^2(t)\rangle \approx \tau_u^2 \langle u^2\rangle n = \tau_u^2\langle u^2\rangle t/\tau_u = \tau_u\langle u^2\rangle t = \frac{k_B T}{\eta}t.$$

Despite the approximations made above, our result is almost right: It is off by only a factor of two. The correct result is

$$\langle x^2(t)\rangle = 2\frac{k_B T}{\eta}t \equiv 2Dt, \qquad (4.16)$$

where

$$D = k_B T/\eta. \qquad (4.17)$$

The quantity D is called the diffusion coefficient. Eq. 4.17, known in the literature as the Einstein-Smoluchowski formula,[9] is another example of a fluctuation-dissipation relationship establishing a connection between temperature, the friction coefficient, and the diffusion coefficient.

Consistent with the expectation for a free particle, its motion is unbound. That is, its mean square displacement grows (linearly) with time. This result is different from the case of a particle moving with a constant velocity, where $\langle x(t)\rangle$ would grow linearly and $\langle x^2(t)\rangle$ would be a quadratic function of time.

Exercise

For a spherical object immersed in a liquid, the friction coefficient η can be estimated using the formula derived by George Gabriel Stokes:

$$\eta = 6\pi\mu a.$$

Here a is the radius of the sphere and μ is the so-called viscosity (or dynamic viscosity) of the liquid. The viscosity of water at room temperature is about $\mu \approx 0.001$ N × s/m^2. Using this formula, estimate the time a single protein molecule, which can be approximated as a sphere with $a = 2$ nm, stays within the observation volume before diffusing away in the experiment shown in Fig.2.1. Assume that the observation volume is also a sphere, whose radius is $r = 1000$ nm.

---

[9] This equation first appeared in 1905 in Albert Einstein's paper on Brownian motion. During the same year, Einstein revealed his special theory of relativity, including his famous relation $E = mc^2$, and published his other famous paper on photoelectric effect, which has established the quantum nature of light and later earned him a Nobel Prize. Marian Smoluchowski discovered Eq. 4.17 independently in 1906.

4.4.2 MOTION IN A SINGLE POTENTIAL WELL

Consider now the motion in a harmonic potential well defined by the equation

$$\tilde{V}(x) = \frac{1}{2}\gamma x^2.$$

Important examples of such motion include the dynamics of a diatomic molecule (in the limit where the interatomic distance remains close to its equilibrium value, as discussed in the beginning of this chapter) and the motion of a bead in an optical trap created by a laser beam, as in Figure 2.5. In the absence of friction ($\eta = 0$) and of the random force, Eq. 4.2 is well known to lead to harmonic oscillations or solutions of the form:

$$x(t) = x_0 \cos(\omega_0 t + \phi),$$

where x_0 and ϕ are constants while

$$\omega_0 = \sqrt{\gamma/m}$$

is the harmonic oscillator frequency. In many single-molecule applications, however, particularly those concerned with biophysical phenomena, the opposite limit of high friction is more pertinent. This will be the limit that we will consider here. More precisely, the following discussion will focus on the so-called overdamped limit, defined as follows: Let τ_r be the characteristic timescale over which $x(t)$ changes by a significant amount. Then dx/dt is of order l_0/τ_r, where l_0 is a characteristic length scale of the system, while d^2x/dt^2 is of order l_0/τ_r^2. The friction force, which is then of order $\eta l_0/\tau_r$, is much greater than the lhs of Eq. 4.2, which is of order ml_0/τ_r^2, if

$$\tau_r \gg \tau_u = m/\eta. \tag{4.18}$$

In the overdamped limit, this condition is true so we can neglect the term containing the acceleration and replace Eq. 4.2 by a 1st order differential equation for $x(t)$:

$$\dot{x} = -(\gamma/\eta)x + R(t)/\eta. \tag{4.19}$$

This equation is formally identical to Eq. 4.12. This observation immediately allows us to identify the characteristic time for the system's relaxation within the potential well:

$$\tau_r = \eta/\gamma.$$

Substituting this into Eq. 4.18 we find that the overdamped limit holds if η/m is much greater than the harmonic oscillator frequency ω_0. By analogy with Eq. 4.12, which describes the relaxation of the velocity of a free particle, we immediately obtain the expression describing the autocorrelation function of the position within the well:

$$\langle x(0)x(t) \rangle = \langle x^2 \rangle e^{-t/\tau_r},$$

where

$$\langle x^2 \rangle = \frac{k_B T}{\gamma} \equiv l_0^2$$

establishes an appropriate characteristic length scale.

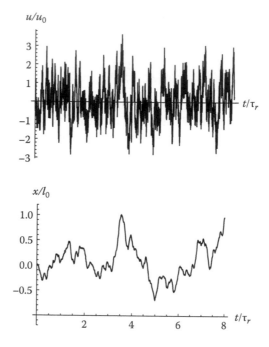

FIGURE 4.5 A typical time dependence of the velocity and position of a particle in a harmonic well. The velocity if measured in units of the root mean square velocity u_0 while the position is normalized by the root mean square deviation l_0 of the system from the bottom of the well.

These findings are illustrated in Figure 4.5, which shows typical computer-generated time dependences of the velocity and the coordinate of a particle in a harmonic potential.[10] Notice that the velocity fluctuates on a much shorter timescale than does the coordinate. This is consistent with the observation that the characteristic timescale for $u(t)$ is τ_u, which—in the overdamped case—is much shorter than the coordinate relaxation timescale τ_r. Thus the system forgets its velocity much quicker than it does its location.

The parameters τ_r and τ_u provide characteristic timescales, over which the position and the velocity attain their respective equilibrium distributions. That is, if we start with a system in some unique initial state described by a position $x(0)$, and velocity $u(0)$, and follow its trajectory for a time t that is substantially longer than both τ_r and τ_u, we will find it in a generic state $(x(t), u(t))$ that has no memory of where the system was at $t = 0$. Moreover, if we repeat this experiment many times, we will find that the values of $x(t)$ and $u(t)$ obey the proper Boltzmann statistics. In the overdamped regime, where $\tau_r \gg \tau_u$, the equilibrium Maxwellian distribution for the velocity (see Appendix B) is established much quicker than the corresponding distribution for the position.

---

[10] Note that, although the parameters of the system were chosen in Fig. 4.5 so as to correspond to an overdamped case, it is necessary to solve the full Langevin equation 4.2 (without neglecting the second derivative term) in order to generate a properly behaved *velocity* dependence $u(t)$.

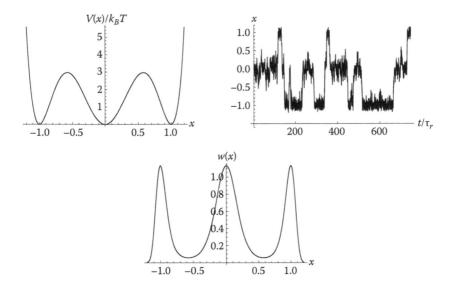

FIGURE 4.6 Motion in a potential that has three wells separated by barriers, which are higher than the thermal energy $k_B T$. The probability distribution $w(x)$ of finding the system at x has three peaks corresponding to the potential minima. The trajectory of the system, $x(t)$, is plotted as a function of time normalized by the position memory time τ_r calculated for the central well. During the time the system is trapped in one well, the distribution of x is close to the single peak corresponding to that well.

4.4.3 MULTIPLE POTENTIAL WELLS

We finally discuss the more complicated case of a potential that has multiple wells, as illustrated in Figs. 4.2 and 4.3. If the depth of each well is substantially greater than $k_B T$, the probability density $w(x) \propto \exp[-\tilde{V}(x)/k_B T]$ of finding the system at x is peaked near the potential minima. This means that the system must be spending most of the time inside the potential wells, close to their minima. Of course, it should also undergo occasional transitions among the wells, but such transitions must be infrequent and/or brief since the Boltzmann statistics prescribes that the fraction of time spent in the high energy regions visited during transitions must be small. Indeed, numerical solution of the Langevin equation for a potential that has several wells (Figure 4.6) shows this kind of behavior.

During the time the system spends inside a single well, the description developed in Section 4.4.2 can be applied. Specifically, we can define a characteristic time, τ_r, that describes the system's position memory loss within the well. Of course, such a quantity is meaningful only if the typical dwell time inside the well, t_{dwell}, is much longer than τ_r. If this is the case then the system will forget the state it was in as it first entered the well long before it exits the well again. Moreover, the distribution of the system's position and momentum during the dwell time should obey a *local* Boltzmann distribution for the well. That is, the shape of the probability density $w(x)$ measured during the time the system dwells in one of the potential wells should be close to $\exp[-\tilde{V}(x)/k_BT]$, which, however, differs from the correct Boltzmann distribution in that only one peak, corresponding to this specific well, is observed.

The physical picture that emerges is quite simple: During its stay within one well, the system behaves as if other wells do not exist. The statistical properties of its trajectory correspond to the local equilibrium within this well. Every once in a while, however, the system wanders too far away from the well minimum and ends up in a neighboring well. This event is followed by thermal equilibration within the new well and so forth. When such a picture is valid, it is often beneficial to adopt a more coarse description of the system, in which, instead of the exact position $x(t)$, one specifies which well is occupied at any given time. This description is especially useful because single-molecule measurements often do not have a sufficient time resolution to measure the instantaneous configuration of a molecule. Rather, they measure molecular properties averaged over a certain time window $\Delta\tau$. For example, if the state of the molecule is measured optically, $\Delta\tau$ must be long enough that the molecule has emitted a sufficient number of photos for the observer to deduce its configuration (see Chapter 7). As a result, the actual quantity accessible experimentally is often the window-averaged position

$$\bar{x}(t) = (\Delta\tau)^{-1} \int_{t-\Delta\tau/2}^{t+\Delta\tau/2} x(t')dt'.$$

Consider now a piece of trajectory $x(t)$ confined to one of the potential wells. If the time resolution $\Delta\tau$ is longer than the characteristic time τ_r, then the above average will simply produce an $\bar{x}(t)$ that is close to the mean position within the well. Such a measurement will therefore lose any information about fluctuations within the well but rather report on which well is occupied. An appropriate description of such a measurement would therefore be in terms of a time-dependent discrete function $i(t)$, where i is the number of the well occupied. When considering its time dependence, it is important to keep in mind that the time itself must be viewed as coarse-grained. Indeed, our description is meaningful only when the time resolution is coarser than the relaxation times with individual wells. This observation will be important in the next chapter, which will provide further connection between the detailed trajectories $x(t)$ and their coarse grained counterparts $i(t)$.

What are the properties of the new, discrete-valued trajectory $i(t)$? Since the underlying dynamics of the system is generated by stochastic equations of motion, $i(t)$ should also be stochastic. That is, it should be described in terms of probabilities of jumping from one state i to another state j. Moreover, the current state of the system, $i(t)$, has no memory of its earlier states because the previous values of $x(t)$ are forgotten over a time comparable to τ_r, which amounts to instant memory loss within our coarse grained picture. This means that future states j depend on the current state i but not on the past states.[11] Let $k_{i \to j}(t)dt$ be the probability to jump from state i to state j between t and $t + dt$ provided that the system is found in i at time t. This probability is a function of i (the state the system is currently in) and j (the state to which the jump may occur). It does not depend on the previously visited states.

---

[11] Such memoryless trajectories $i(t)$ are known as *Markov processes*.

Moreover, if the external conditions (the temperature or the potential to which the system is subjected) do not change with time, then $k_{i \to j}$ should not depend on time t. Let w_i be the probability to find the system in state i at time t. After a short interval of time dt, the probability may change because the system may jump into a new state j or it can jump from another state j to i. Specifically, we can write

$$dw_i = -\sum_{j \neq i} k_{i \to j} w_i dt + \sum_{j \neq i} k_{j \to i} w_j dt$$

or

$$\frac{dw_i}{dt} = -\sum_{j \neq i} k_{i \to j} w_i + \sum_{j \neq i} k_{j \to i} w_j. \tag{4.20}$$

We have arrived at the *master equation*, a simple, phenomenological description of a system's dynamics as hopping among discrete states corresponding to the system's energy minima. We have already seen this equation, for a case of two states we were referring to as molecular species A and B, in the preceding chapter. Because of their remarkable simplicity, master equations are widely used. Chemical kinetics discussed in the previous chapter is just one example. Another example involves the emission of photons by a single molecule: When the index i labels different energy levels of a molecule, the resulting master equation can be used to describe the process through which a molecule emits or absorbs photons, each individual emission or absorption event corresponding to a jump to a lower or higher energy level. This case will be further discussed in Chapter 7.

The rate coefficients $k_{i \to j}$ appearing in the master equation satisfy several important constraints. For a system that obeys Boltzmann statistics, the stationary solution of Eq. 4.20 must coincide with the Boltzmann distribution. This means that, if we take the $t \to \infty$ limit in Eq. 4.20, the resulting probability of occupying the well i must agree with the Boltzmann probability of finding the system in the well i, which is equal to

$$w_i(\infty) = q_i/q = q^{-1} \int_{x \approx x_i} dx \exp(-\beta \tilde{V}(x)). \tag{4.21}$$

Here q_i is the partition function associated with the state i and $q = \sum_i q_i$ is the total partition function (cf. Appendix B). The integration in the above formula is performed over the conformations corresponding to the potential well i. Because we have assumed that the depth of the well is much greater than $k_B T$, this integral is dominated by the values of x that are close to the well bottom x_i and so the precise integration boundaries are unimportant. Substituting Eq. 4.21 into Eq. 4.20, we find

$$-\sum_{j \neq i} k_{i \to j} w_i(\infty) + \sum_{j \neq i} k_{j \to i} w_j(\infty) = 0. \tag{4.22}$$

This condition is automatically satisfied if the rate coefficients all satisfy the *detailed balance condition*

$$k_{i \to j} w_i(\infty) = k_{j \to i} w_j(\infty) \tag{4.23}$$

and

$$k_{i \to j}/k_{j \to i} = q_j/q_i. \tag{4.24}$$

The second of these two equations indicates that the ratio of the two rate coefficients, which are quantities that characterize the dynamics, is actually constrained by the equilibrium statistics of the system. Indeed, microscopic rate theory developed in the next chapter will show the validity of this condition and will shed further light on its microscopic origins. Detailed balance guarantees that the master equation Eq. 4.20 has a stationary solution (i.e., one for which the state occupation probabilities do not change with time, $dw_i/dt = 0$) and that this solution coincides with the Boltzmann distribution. Further consequences of having detailed balance are discussed in the next section.

It should however be noted that detailed balance is a sufficient but not a *necessary* condition for the existence of a stationary solution of the master equation. Consider, for example, the system composed of 3 states, $i = 1, 2, 3$, with the following rate coefficients:

$$k_{1\to2} = k_{2\to3} = k_{3\to1} = k$$

$$k_{2\to1} = k_{3\to2} = k_{1\to3} = 0.$$

Substituting these into Eq. 4.20, one obtains

$$dw_1/dt = kw_3 - kw_1$$

$$dw_2/dt = kw_1 - kw_2$$

$$dw_3/dt = kw_2 - kw_3.$$

These equations describe a stochastic process, in which the system cycles through the three states:

$$\cdots \to 1 \to 2 \to 3 \to 1 \to 2 \to \cdots. \tag{4.25}$$

That is, the sequence of the states in itself is deterministic while the transition times are random. The master equation has an obvious stationary solution:

$$w_1 = w_2 = w_3 = 1/3.$$

This solution, however, violates the detailed balance condition of Eq. 4.23. Moreover, this master equation generates unidirectional motion that violates the time-reversal symmetry: The time-reversed process would cycle through the same states backwards. Such behavior, therefore, would be inconsistent with the assumption that the underlying physical model describing the system's trajectories is one described by Newtonian dynamics, as Newton's equations of motion, Eq. 4.1, are time-reversible (see Chapter 6 for a further discussion of time reversibility of Newtonian and Langevin dynamics).

Do not, however, rush to discard the above kinetic scheme as physical nonsense! It turns out that master equations violating detailed balance commonly arise as an approximate description of *nonequilibrium* processes. For example, molecular machines in our body are capable of generating unidirectional or rotary motion. This motion involves consumption of "fuel" molecules. The apparent violation of detailed balance and lack of time reversibility is then a consequence of incomplete description of the system. Should the "fuel" be included explicitly, the resulting process will be time-reversible, at least when observed over a sufficiently long timescale. We will discuss models violating detailed balance in Chapter 10.

Coarse-grained master equations enormously simplify the description of any molecular system. We should, however, keep in mind that such a description is not always possible. Unlike the situation depicted in Fig. 4.6, real molecular potentials do not always consist of deep wells separated by large barriers. Many molecular systems undergo motion on relatively flat energy landscapes. This is especially true for polymers, long chain molecules that can adopt multiple configurations of comparable energy. The appropriate level of description (quantum mechanics, classical mechanics, stochastic Langevin dynamics, or master equation) should be chosen on a case-by-case basis.

4.5 FURTHER DISCUSSION: SOME MATHEMATICAL PROPERTIES OF THE MASTER EQUATION

This section describes some of the important mathematical properties of the master equation, Eq. 4.20. This equation describes the time evolution of a stochastic system that can occupy any of its N states, which are labeled by indices $i, j = 1, 2, \ldots, N$. The probability of finding the system in state i at time t is given by $w_i(t)$. The properties of the system are completely specified by the initial conditions $w_i(0)$ and by the set of rate coefficients $k_{i \to j}$ between each pair of states. Of course, not all of the rate coefficients have to be nonzero. I will, however, assume that all the states are "connected" in that, for any pair of states i and j, there is a sequence of intermediate states, $i_1, i_2, \ldots, i_M$, such that the rate coefficients $k_{i \to i_1}, k_{i_1 \to i_2}, \ldots, k_{i_M \to j}$ are nonzero. In other words, it is possible to arrive in any state j, having started from any state i, either directly or through a sequence of intermediate states. If this is not the case then it makes sense to break all the states into the subsets of connected states. The dynamics within each subset is then completely decoupled from that of all other subsets and so each subset can be studied separately.

I will further assume that the rate coefficients satisfy the detailed balance condition:

$$k_{i \to j} w_i^{(eq)} = k_{j \to i} w_j^{(eq)}, \tag{4.26}$$

where $w_i^{(eq)}$ is the equilibrium population of state i. The master equation can be rewritten in a matrix form:

$$\frac{d\mathbf{w}}{dt} = -\mathbf{K}\mathbf{w}, \tag{4.27}$$

where $\mathbf{w}$ is a column vector with components w_i and $\mathbf{K}$ is a square matrix, whose elements are given by $K_{ij} = -k_{j \to i}$ for $j \neq i$ and $K_{ii} = \sum_{j \neq i} k_{i \to j}$. The solution of this equation can be written as

$$\mathbf{w}(t) = \mathbf{T}(t)\mathbf{w}(0), \tag{4.28}$$

where the transition matrix $\mathbf{T}$ is equal to the matrix exponent

$$\mathbf{T}(t) = \exp(-\mathbf{K}t). \tag{4.29}$$

Consider the probability $w_i(t)$ to find the system in state i at time t. According to Eq. 4.28, this can be written as

$$w_i(t) = \sum_j T_{ij}(t) w_j(0),$$

allowing us to interpret the matrix elements $T_{ij}(t)$ as the conditional probabilities to find the system in a state i at time t provided that it was in state j at time 0.

If the transition matrix $\mathbf{T}$ is known, the rate coefficients of the underlying master equation can be recovered by using the following relation:

$$k_{i \to j} = \lim_{t \to 0+} dT_{ji}(t)/dt. \qquad (4.30)$$

This result is readily proven by taking the time derivative of the short-time approximation to Eq. 4.29,

$$\mathbf{T} \approx \mathbf{1} - \mathbf{K}t,$$

where $\mathbf{1}$ denotes the identity matrix. Eq. 4.30 formalizes our definition of a rate coefficient as transition probability per unit time.

Introducing the eigenvectors $\mathbf{u}_\alpha$ and eigenvalues λ_α of the matrix $\mathbf{K}$, defined by the equation

$$\mathbf{K} \mathbf{u}_\alpha = \lambda_\alpha \mathbf{u}_\alpha, \qquad (4.31)$$

we can also write the solution of the master equation as a spectral expansion,

$$\mathbf{w}(t) = \sum_\alpha c_\alpha \exp(-\lambda_\alpha t) \mathbf{u}_\alpha, \qquad (4.32)$$

where the coefficients c_α can be calculated if the vector $\mathbf{w}(0)$ of the initial probabilities is known. One important property of the eigenvalues λ_α is that they are real numbers. To show this, consider one of the eigenvalues λ and the corresponding eigenvector $\mathbf{u}$ and rewrite Eq. 4.31 explicitly as follows:

$$\sum_j K_{ij} u_j = \lambda u_i. \qquad (4.33)$$

Now define a rescaled matrix

$$\tilde{K}_{ij} = \frac{K_{ij} w_j^{(eq)}}{\sqrt{w_i^{(eq)} w_j^{(eq)}}}$$

and rescaled eigenvector $v_j = u_j / \sqrt{w_j^{(eq)}}$. In terms of those, the above eigenvalue equation can be rewritten as

$$\sum_j \tilde{K}_{ij} v_j = \lambda v_i.$$

Therefore, λ is also an eigenvalue of the rescaled matrix $\tilde{\mathbf{K}}$. In view of the detailed balance condition, Eq. 4.26, the rescaled matrix $\tilde{\mathbf{K}}$ is symmetric and so the eigenvalue

λ is a real number. This, in particular, implies that our master equation cannot have oscillatory solutions.

Further physical constraints on the eigenvalues λ_α come from two physical requirements. Firstly, the probabilities w_i are nonnegative numbers that cannot exceed one. Thus negative values λ_α are unphysical, as, according to Eq. 4.32, they would lead to exponentially growing probabilities. Secondly, all λ_α's cannot be positive because, according to Eq. 4.32, this would make all probabilities vanish in the limit $t \to \infty$, a result that would contradict the assertion that the system is always found in one of the N states and so the probabilities w_i must add up to one at any time t.[12] Instead, in the limit $t \to \infty$ the populations of states should be approaching their equilibrium values, i.e., $w_i(\infty) = w_i^{(eq)}$. Therefore, one of the λ_α's, say the one corresponding to $\alpha = 0$, must be equal to zero, $\lambda_0 = 0$. The corresponding eigenvector $\mathbf{u}_0$ is, to within a constant normalization factor, given by

$$\mathbf{u}_0 = (w_1^{(eq)}, w_2^{(eq)}, \ldots, w_N^{(eq)})^T.$$

Indeed, substituting $\mathbf{w} = \mathbf{u}_0$ into the rhs of Eq. 4.27 results in the identity $d\mathbf{w}/dt = 0$, as anticipated for equilibrium probabilities.

Finally, consider any quantity (e.g., position, etc.) x that takes on discrete values $x = x_1, x_2, \ldots, x_N$ when the system is, respectively, in states $i = 1, 2, \ldots, N$. The trajectory $x(t)$ is therefore a piecewise function that remains constant as long as the system remains in the same state. Computation of the correlation functions of the form $\langle x(0)x(t) \rangle$ is a commonly encountered problem. For a system obeying the above master equation, this correlation function can be calculated as

$$\langle x(0)x(t) \rangle = \sum_{i,j} x_j T_{ji}(t) x_i w_i^{(eq)} = \mathbf{x}^T e^{-\mathbf{K}t} \tilde{\mathbf{x}},$$

where $\tilde{\mathbf{x}}$ is the column-vector with the components $(x_1 w_1^{(eq)}, x_2 w_2^{(eq)}, \ldots, x_N w_N^{(eq)})$.

Exercise

A single molecule can be found in three states, A, B, and C, and obeys the following reaction scheme:

$$A \underset{k_{B \to A}}{\overset{k_{A \to B}}{\rightleftharpoons}} B \underset{k_{C \to B}}{\overset{k_{B \to C}}{\rightleftharpoons}} C.$$

The molecule emits light (intensity $I = I_0$) when in the state A and is dark ($I = 0$) in the states B and C. What is the resulting autocorrelation function $\langle I(0)I(t) \rangle$ of the light intensity?

---

[12] The fact that the sum of the probabilities $W(t) = \sum_i w_i(t)$ is conserved is readily proven by considering its time derivative $dW/dt = \sum_i dw_i/dt$. To calculate dW/dt we thus simply sum the rhs of Eq. 4.20 over i, which gives zero.

4.6   FURTHER DISCUSSION: HOW DOES A MOLECULE "KNOW" ITS OWN ENTROPY?

The discussion of this chapter suggests that ergodicity of a system, from a practical standpoint, is closely connected to the efficiency with which the system can explore its conformational space. Here, we will try to quantify this efficiency by considering the very simple, textbook case of an ideal gas confined to a box. The molecules of the gas do not interact, except for an occasional collision with the walls of the container or with each other.

In the spirit of this book, let us focus on just one molecule of the gas. We would like to estimate the time before we are sure it has explored all the space available to it. To make our task more precise, it is convenient to adopt a quantum-mechanical view, according to which a molecule confined to a finite volume can be in any of its quantum states, whose energies take on discrete values. The goal is then to estimate the time it takes the molecule to visit all of the accessible states. If we ignore the molecule's rotational and vibrational degrees of freedom (or if we are dealing with a monoatomic gas), then we can think of this molecule as a particle in a box, another textbook problem in physics. The allowed energies for a quantum particle of mass m that is confined to a three-dimensional box with the dimensions $L_x \times L_y \times L_z$ are given by the formula:

$$E_{n_x,n_y,n_z} = \frac{\hbar^2\pi^2 n_x^2}{2mL_x^2} + \frac{\hbar^2\pi^2 n_y^2}{2mL_y^2} + \frac{\hbar^2\pi^2 n_z^2}{2mL_z^2}, \tag{4.34}$$

where n_x, n_y, and n_z are positive integer numbers and $\hbar$ is Planck's constant. Assuming that the molecule is in thermal equilibrium with the other gas molecules and/or with the walls of the container, the probability of finding the molecule in a state with the energy E_{n_x,n_y,n_z} is proportional to the Boltzmann factor $\exp(-\frac{E_{n_x,n_y,n_z}}{k_BT})$. This means that the states with energies such that $E_{n_x,n_y,n_z} \gg k_BT$ are unlikely to be occupied. We will ignore those thermally inaccessible states and focus on the states whose energies are not much higher than, say, a few k_BT. Let us, for example, estimate the number of states W with the energy less than k_BT. Using Eq. 4.34, the condition $E_{n_x,n_y,n_z} < k_BT$ amounts to

$$n_x^2 + n_y^2 + n_z^2 = n^2 < \frac{2mL^2 k_BT}{\hbar^2\pi^2} \tag{4.35}$$

where we have assumed the box to be a cube of volume L^3. The points with the coordinates $\mathbf{n} = (n_x, n_y, n_z)$ fill a sphere of radius n. The number of such points can therefore be estimated as the volume of this sphere,

$$W \approx 4\pi n^3/3 = (4\pi/3)L^3\left(\frac{2mk_BT}{\hbar^2\pi^2}\right)^{3/2}. \tag{4.36}$$

Before we discuss the numerical value of W, it is instructive to derive this result in a different way. In classical mechanics, "the state" of a particle is defined by its momentum (p_x, p_y, p_z) and position (x, y, z) vectors. We can think of such states as vectors (x, y, z, p_x, p_y, p_z) occupying a six-dimensional space (referred to as the phase space). Since the coordinates and the momenta are continuous variables, the number of such states is infinite. The quantum mechanical uncertainty principle, however, imposes a fundamental limit on the resolution, with which we can define a state (see also Appendix B). It states that the uncertainties, with which one can measure the position x and the momentum p_x, must satisfy the following relationship:

$$\Delta p_x \Delta x \sim \hbar. \tag{4.37}$$

We can then think of the phase space as consisting of cells of the volume approximately equal to

$$\Delta p_x \Delta x \Delta p_y \Delta y \Delta p_z \Delta z \sim \hbar^3.$$

The number of such cells is the number of states we seek. This number can be estimated as

$$W = \frac{\text{the volume of phase space}}{\text{the volume of the cell}}. \tag{4.38}$$

The volume of the phase space that corresponds to all states with the energy below $k_B T$ can be estimated as

$$\int_{\frac{p_x^2}{2m}+\frac{p_y^2}{2m}+\frac{p_z^2}{2m}<k_B T} dx\,dy\,dz\,dp_x\,dp_y\,dp_z = L^3 \int_{\frac{p_x^2}{2m}+\frac{p_y^2}{2m}+\frac{p_z^2}{2m}<k_B T} dp_x\,dp_y\,dp_z. \tag{4.39}$$

The integration over the momenta, again, gives the volume of a sphere. It is easy to see that Eqs.4.38 and 4.39 lead to a result which, to within a numerical factor, is the same as Eq. 4.36. Since our estimates are quite crude, we should not worry about numerical factors. Finally, notice that our result for W can also be written as:

$$W \sim \left(\frac{L}{\lambda_{dB}}\right)^3. \tag{4.40}$$

Here

$$\lambda_{dB} = 2\pi\hbar/(mu) \sim \hbar/(mk_B T)^{1/2} \tag{4.41}$$

is what is called the molecule's thermal de Broglie wavelength and u is its typical thermal velocity estimated from the requirement that the molecule's kinetic energy, $mu^2/2$, is of order $k_B T/2$. To be specific, consider an oxygen molecule at $T = 300K$. Its mass can be calculated as $m = M_{O_2}/N_a$, where $M_{O_2} = 0.032kg/mol$ is the molar mass of oxygen and $N_a \approx 6.02 10^{23}$ is Avogadro's number, and so its thermal velocity is roughly $u \approx 300m/s$. The corresponding de Broglie wavelength, λ_{dB}, is in a sub-Angstrom range. We can therefore estimate W as the number of tiny cells of volume $\sim \lambda_{dB}^3$ that can be placed inside the volume L^3. For a box with $L = 1m$, Eq. 4.35 gives

$$W \approx 10^{33}. \tag{4.42}$$

This is a huge number! It would be even larger if we took the rotational and vibrational degrees of freedom into consideration. To estimate how much time it will take our molecule to visit all of these states, we first need to know how much time it spends in one state before going to another. Suppose every time it undergoes a collision with the walls of the container or with another molecule it jumps into another quantum state. If the gas is extremely dilute and the molecule collides with the walls more frequently than it does with other molecules then the time between collisions with the container walls can be estimated as

$$\tau_{wall} \sim L/u$$

where u is the typical thermal velocity estimated above. This gives $\tau_{wall} \sim 0.0003s$. This provides an upper bound on the time between the collisions since collisions with the other molecules have so far been neglected. To estimate the typical time between intermolecular collisions, imagine tracking the molecule's position during some time interval t. It will collide with every other molecule that came within a certain distance D from the molecule's path, where D is roughly comparable to the size of the molecule. If the length of the path is $l = ut$ then the number of the molecules that act as obstacles is simply $\pi D^2 l\rho$, where ρ is the number of molecules per unit volume. The average distance traveled between two collisions is

$$\lambda = l/(\pi D^2 l\rho) = (\pi D^2 \rho)^{-1} \tag{4.43}$$

and the time between collisions is

$$\tau \approx \lambda/u = (\pi D^2 \rho u)^{-1}.$$

Air, at room temperature, has $\rho \approx 2.7 \times 10^{25}$ molecules per cubic meter. We do not exactly know what D is but it should be comparable to a typical molecular size (cf. Chapters 1-2). Let us say $D = 3\mathring{A} = 3 \times 10^{-10}m$. This gives $\tau \approx 4 \times 10^{-10}s$. It therefore takes less than a nanosecond between two successive collisions. The time it takes to explore all of the W states can be roughly estimated as

$$t_{explore} = \tau W = 4 \times 10^{23}s. \tag{4.44}$$

This result underestimates the true value of $t_{explore}$ because it ignores the rotational and vibrational quantum states of the molecule and because each collision is assumed to lead to a new state. Nevertheless, our estimate exceeds the age of our Universe, which happens to be about $t_{Universe} \approx 4 \times 10^{17}s$, by six orders of magnitude! That is, on any practically relevant timescale, a single molecule only explores an infinitesimal fraction of its states.

Given that the molecule will never visit the overwhelming majority of its accessible states, can we think of it as behaving ergodically? Consider Figure 4.7, which illustrates a trajectory of an ideal gas molecule. The phase space of the system is divided into small cells representing states. While this space is actually six-dimensional we only have two dimensions available for illustration purposes. In our depiction of the phase space we have also ignored the lack of hard-wall boundaries in the momentum space: There is no strict threshold value that the molecule's momentum cannot exceed but, rather, high values of momenta (and—correspondingly—energy) are suppressed by the Boltzmann statistics. The most significant way, in which this figure diverges from reality, is that the cells are shown to be too large. In fact, there are just 2500 cells shown in Fig. 4.7, in contrast to the astronomically large number estimated from Eq. 4.42.

Although the number of the cells shown in Fig. 4.7 is not very large, it nevertheless appears that a relatively coarse representation of the molecule's trajectory, which

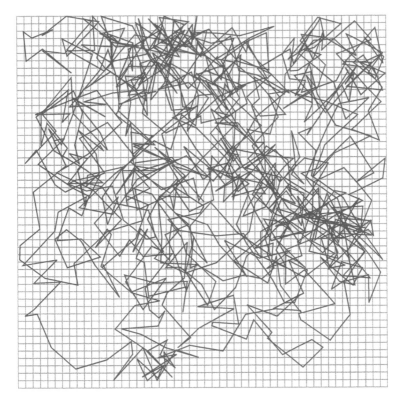

FIGURE 4.7 A trajectory of a molecule confined to a phase-space square. The square is divided into 2500 cells. Despite the fact that the molecule has not visited every one of them, even this relatively short piece of trajectory will tell us that the mean position of the molecule is close to the middle of the square.

records what cell is visited at any given time and which uses a reasonably small number of cells (perhaps more than 50×50 cells shown in the picture, but still small enough), may suffice for many practical purposes, just like the resolution of your computer screen is likely sufficient for viewing your vacation photos. Moreover, given a sufficiently long trajectory (somewhat longer than that shown in Figure 4.7), one can accurately estimate average properties of the molecule (e.g., its average position or the average energy) even if this trajectory does not cross every cell in the figure. Even the relatively short trajectory shown in Fig. 4.7 will, for example, tell us with reasonably good accuracy that the average position of the molecule is right in the middle of the square. Moreover, dynamical properties such as the molecule's diffusion coefficient can also be estimated.

It is instructive to estimate the length of the trajectory required in order for such estimates to be accurate and to compare it with $t_{explore}$. To this end, we think of the molecule's trajectory as a random walk with an average step length λ estimated from Eq. 4.43. We know (cf. Appendix A) that, for a random walk in free space, the distance

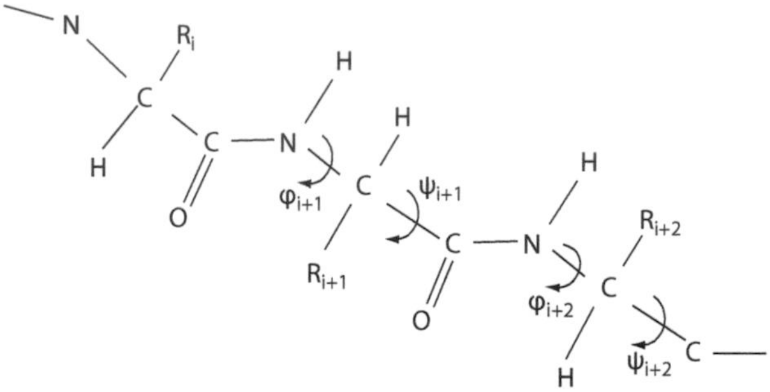

FIGURE 4.8 Levinthal paradox: Proteins are long chain molecules, which assume well-defined structures under physiological conditions. The conformation of the molecule can be described by a set of angles, $\phi_1, \psi_1, \phi_2, \psi_2, \ldots, \phi_N, \psi_N$. If, for example, each of these angles can take on 3 possible values and if $N = 50$, this results in $W = 9^{50}$, an astronomical number of possible configurations, out of which only one (or may be a relatively small subset) correspond to the correct, native configuration adopted in our body.

r traveled after k steps satisfies the relationship

$$\langle r^2 \rangle = \lambda^2 k. \tag{4.45}$$

Therefore it takes the molecule roughly $k = L^2/\lambda^2$ steps or

$$\tau_{diff} \approx (\lambda/u)k = L^2/(\lambda u) \approx 2.5 \times 10^4 s \approx 7 \text{ hours}$$

to diffuse across the length of the container. It appears sensible that, in order for the trajectory to uniformly fill the available space, the length of the trajectory must be longer than τ_{diff}. Indeed, τ_{diff} sets the characteristic timescale that determines whether or not the motion of a molecule confined to a box will appear ergodic. If the length t of the trajectory is such that $t \ll \tau_{diff}$ then the trajectory will only explore only part of the box. If, in contrast, $t \gg \tau_{diff}$, then average properties of the molecule (such as its mean position) can be reliably estimated from this trajectory. Moreover, they will coincide with the same properties calculated as ensemble averages performed over all molecules of the gas. Note that, in the above calculation, we have conveniently forgotten about exploration of the momentum space and limited our attention to the random walk in the xyz space. The characteristic timescale τ_p for the exploration of the momentum space is different. It turns out that, in most cases, this timescale is considerably shorter than τ_{diff}. In fact, τ_p is related to the velocity memory time τ_u introduced earlier in this chapter. Usually, a molecule forgets its velocity much sooner than it does its position (cf. Figs. 4.4 and 4.5), as even a single collision can drastically change the molecule's velocity but not its location in space.

The timescale τ_{diff} is much shorter than the astronomically long timescale $\tau_{explore}$ required to exhaustively sample all the quantum states available to the molecule. Since all time averaged properties considered so far are expected to converge to their correct values as long as $t \gg \tau_{diff}$, it seems that molecules' failure to sample all the available states exhaustively does not have any dramatic physical consequences. Even though, on the very fine scales set by the quantum mechanical phase space volume cells, the molecule fails to visit many (in fact most of) such cells, it nevertheless visits the "typical" cells and so any statistically averaged properties of the molecule are correct, despite very incomplete sampling. Only when much higher resolution is required, e.g., if we want to zoom in on a single state or a small group of states that such incompleteness of sampling may have consequences. One example when the issue of incomplete sampling was brought to the spotlight is the famous paradox due to Cyrus Levinthal. Levinthal pointed out—using arguments similar to those presented above—that the time it takes a random molecular chain to find the correct, "native" configuration (i.e., the protein structure found in the cell) would be longer than the age of the Universe, provided that it searches for the native state randomly (Figure 4.8). Indeed, in this case the molecule needs to find a single "target state," or perhaps a small subset of target states, out of the immense number W of possibilities. Levinthal's paradox has intrigued many physicists and its mechanistic resolution still remains to be the subject of some debate. Nevertheless, at least in principle, it is clear that there is no reason for the search to be entirely random, as it is biased by the energetics of the problem. Indeed, the same argument could be applied to, say, an apple that has fallen off a tree. If each molecule of the apple were to randomly search for the final state corresponding to the fruit lying on the ground then, considering the very large number of molecules in the apple, we could conclude that the time for the apple to reach the ground would be longer than the age of the Universe. It is clear in this case that the cohesive interactions among the apple's molecules are essential for overcoming Levinthal's paradox: Remove them and, in perfect accord with Professor Levinthal, the apple will become vaporized never ending up on the ground.

Although the above arguments suggest that average molecular properties converge to their correct, thermodynamic values over a time that is comparable to τ_{diff}, there is no guarantee that some of them would not take much longer to converge. One disconcerting example of such a property is the total number W of quantum states itself. Indeed, W is directly related to the system's entropy S through the famous formula due to Boltzmann:

$$S = k_B T \ln W \qquad (4.46)$$

The entropy of a material can be measured experimentally by placing it in a calorimeter, a device that measures how much heat q you need to increase the material's temperature by a certain amount. If the material, initially at a temperature $T = T_0$, is heated slowly until it achieves some higher temperature T_1, then the change of its entropy is given by

$$S_1 - S_0 = \int_{T_0}^{T_1} dq/T. \qquad (4.47)$$

If T_0 is very close to zero then we know that the system is in the ground state and so the number of accessible states W is close to 1. This means that S_0 is close to zero. By heating the system to, say, room temperature (where W is vary large) and using Eq. 4.47 we have thus measured the entropy of the system, from which we can deduce its number of states W. Of course, doing calorimetry on a single molecule could be difficult but we do not have to: Instead, we could measure the number of states of the entire gas, from which, if desired, the number of states of its single molecule can also be deduced. What value do we get for W? Since the actual number of states

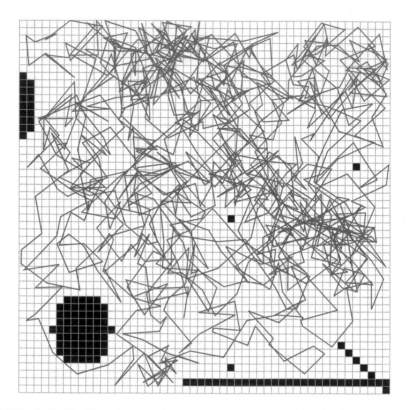

FIGURE 4.9 By blocking off parts of space that have not been visited by the molecule (cells shown in black), one reduces the entropy of the system.

accessed by the system depends on the observation time (unless the experiment has been performed over a period much longer than the age of the Universe), we might expect that the outcome of the measurement would depend on time. This is not so: Provided that the measurement is performed slowly enough,[13] the outcome of the measurement will be the *total* number of states! How can this be? How does our system know how many states it has if it will never visit most of them?

Our paradox can be rephrased somewhat differently. Suppose that during the observation time the molecule has actually visited w states, where w is astronomically smaller than the actual number of states W. Suppose now that we design a different system, which has exactly the same w states but a smaller total number of states, W', because $W - W'$ states have been "blocked off." In our particle-in-a-box example, states can be blocked off by simply filling some parts of the container with obstacles, shown in Fig. 4.9 in black. Because our molecule has never visited the forbidden space, it cannot tell the difference between the two situations. The two systems appear

---

[13] The precise meaning of "slowly enough" depends on the physics of the system. The point is that such sufficiently slow heating can often be attained in a laboratory without having to wait for an excessively long time comparable to the age of the Universe.

FIGURE 4.10 It is impossible to block off $\phi = 20\%$ of the available volume with obstacles of any size such that the molecule will not run into an obstacle over a time comparable to τ_{diff}.

to behave identically yet their actual entropies are different. We thus appear to have tricked our molecule (or the observer) into miscalculating the molecule's own entropy.

One problem with the above proposal is that, before closing off some parts of the container, we need to make sure those are ones the molecule has not yet visited. That is, we need to know in advance which states our molecule would have visited. This is unrealistic. Instead, let us strategically place some obstacles inside the container in the hope that the molecule will not notice them. In order to significantly change the entropy, the number of forbidden states, $W - W'$, must not be too small. Indeed, the change of entropy caused by the obstacles is $\Delta S = k_B \ln(W'/W)$ so if W'/W is close to one this change would negligible. Let us say we have blocked off 20% of the initial volume and, since the number of states is proportional to the volume (cf. Eq. 4.36), this has reduced the number of states by 20%. We seem to be able to afford doing so since the actual number of states w visited by the molecule is many, many orders of magnitude smaller than either W or W'. We can have obstacles of any size and can place them anywhere we want. We could, for example, install a wall inside the container such that the actual volume available to the molecule is $0.8L^3$. This, however, seems to be a bad choice since it will take the molecule a time of order of τ_{diff} to reach the wall and thus to discover our conspiracy. Can we do better? We can try to place many obstacles of a smaller size, as shown in Fig. 4.10. To attain the same value of W', we have to ensure that the total volume occupied by the obstacles remains the same. If the average (linear) size of an obstacle is a then we must require that $\phi = a^3 \rho_{obst}$, where ρ_{obst} is the number of obstacles per unit volume, is constant. The quantity ϕ is the fraction of the total volume occupied by the obstacles. Now, following the same arguments that have led us to Eq. 4.43, we can estimate the "mean free path" traveled by the molecule between hitting two obstacles:

$$\lambda_{obst} \propto \frac{1}{\rho_{obst}a^2} \propto \frac{a}{\rho_{obst}a^3} \propto \frac{a}{\phi}.$$

Since the volume fraction ϕ is fixed, our equation suggests that the smaller the obstacles, the more frequently our molecule will collide with them! We have thus failed to trick the molecule into moving as if it is in free space. It will take a time comparable to τ_{diff} for it to discover that a fraction ϕ of the space is unavailable. In other words, it is unnecessary to visit all the available states to discover that 20% of them are blocked. As a result, the molecule "knows," at least with a very high probability, how many states it has without ever visiting most of them.

5 Microscopic View of the Rate of a Chemical Reaction: A Single-Molecule Perspective

It must therefore be assumed, to be consistent, that the other actual reacting substance is not cane sugar, since the amount of sugar does not change with temperature, but is another hypothetical substance which is regenerated from cane sugar as soon as it is removed through the inversion.This hypothetical substance, which we call "active cane sugar," must rapidly increase in quantity with increasing temperature (by about 12 percent per degree) at the expense of the ordinary "inactive" cane sugar. It must be formed from cane sugar at the expense of heat (of q calories).

Svante Arrhenius, *On the reaction velocity of the inversion of cane sugar by acids*

5.1 FROM MICROSCOPIC DYNAMICS TO RATE COEFFICIENTS

In the previous chapter, we argued that the chemical kinetics equations describing time evolution of different chemical species arise naturally from the microscopic motion governed by the underlying potential energy landscape. Deep wells on the potential energy surface (deeper than the thermal energy $k_B T$) represent molecular species. While most of the time is spent within those wells, a jump from one well to another takes place every once in a while. Such jumps are what we refer to as chemical reactions. Adopting this view, we no longer care about the exact molecular configuration (i.e., the exact position of the system on the energy landscape) but rather characterize the system's state by specifying the well it occupies at any given time. This, in particular, implies that the system always occupies some well, an assumption that can be valid only if the duration of the jump itself is much shorter than the typical time spent inside a well and so the time spent in transit between wells can be neglected. While such a picture often provides a very good description of many molecular phenomena, the transit time is also a very interesting quantity to consider: It will be studied in the next chapter.

Within this simplified view, we have further claimed that the stochastic process, through which the system jumps from one state (or chemical species) to another, can

59

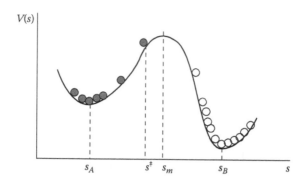

FIGURE 5.1 A simple double-well model of a chemical reaction. Molecule's configurations lying to the left of $s = s^{\ddagger}$ correspond to the reactant state A. Those are shown as dark circles. Molecules with $s > s^{\ddagger}$ (shown as empty circles) correspond to the product B. Most molecules found within each well gather in vicinity of its minimum. To study the population kinetics in this double-well potential, one may create a noneqilibrium situation by removing (or ignoring), at $t = 0$, all molecules B. Of course, at a later time, some of the molecules will be found to have crossed over to B.

be described in terms of a set of rate coefficients, $k_{A \to B}$, $k_{B \to A}$, etc., which are related to the probabilities, per unit time, for making transitions from A to B, B to A, etc. Our story is, however, incomplete because we still do not know how, given the underlying energy landscape, to determine the resulting rate coefficients for all possible pairs of states. This chapter describes how to do so.

Let us concentrate on a pair of chemical species, A and B, and pretend that the configuration of a molecule can be described by a single coordinate s (multidimensional generalizations will be described later in this chapter). The potential energy, $V(s)$, as a function of s, has a double-well shape shown in Figure 5.1, with the left and the right wells, respectively, representing A and B. Molecular trajectories $s(t)$ are generally stochastic because of the molecule's interactions with its surroundings. For example, $s(t)$ may be governed by a Langevin equation introduced in the previous chapter. Whatever the precise model that describes $s(t)$, we will assume that, if we observe our system for a sufficiently long time, we will find that the probability that it visits a point s with a momentum p is proportional to the Boltzmann factor $\exp[-\beta V(s) - \beta p^2/(2m)]$, where, as usual, $\beta = (k_B T)^{-1}$ (see Appendix B).

Our goal is to predict the rate coefficient $k_{A \to B}$ given the potential $V(s)$ and the underlying model that describes the time evolution of the trajectories $s(t)$. Although we will see below that simple analytical approximations exist in certain cases, solving this problem generally requires computation. Suppose we have a computer program that solves the molecule's equations of motion and generates $s(t)$. How can we use it to compute $k_{A \to B}$? At first glance, this task appears quite simple. Indeed, two computer experiments, mimicking real experiments, can be readily designed, one corresponding to bulk chemistry and the other to a single molecule experiment. In the first experiment, one creates, at $t = 0$, an initial ensemble of N trajectories with different initial coordinates $s_i(0)$ and momenta $p_i(0)$, where $i = 1, \ldots, N$. These initial parameters are random numbers drawn from the Boltzmann distribution such that the probability of starting with an initial momentum $p(0)$ from an initial configuration $s(0)$ is proportional to the Boltzmann factor,

$$\exp\{-\beta[V(s(0)) + p^2(0)/2m]\}. \tag{5.1}$$

Now we let each ensemble member evolve in time independently. We could, for example, have N computers, each computing the coordinate $s_i(t)$ and the momentum $p_i(t)$, as a function of time, for one of the N molecules. If, at some later time t, we examine these values, we will discover that our ensemble still obeys the Boltzmann statistics because the system in equilibrium stays in equilibrium as the time evolves.

This is, of course, not exactly what we want as our goal is to study kinetics rather than to observe the system's equilibrium. Nevertheless, it is straightforward to modify our procedure such that the initial condition is no longer an equilibrium one. First of all, let us define, more precisely, what "A" and "B" means. We define any molecular configuration that satisfies the inequality $s < s^{\ddagger}$, where $s^{\ddagger}$ is some point in between A and B, as corresponding to A. Otherwise it is B. For the purpose of counting the molecules of each type, N_A and N_B, the precise choice of the boundary $s^{\ddagger}$ does not matter. Indeed, as the probability of finding a molecule at s is proportional to $\exp[-V(s)/k_BT]$, the majority of the molecules situated within each well gather within a few k_BT from the well minimum. Therefore any $s^{\ddagger}$ placed between the two potential wells such that it is not too close to either minimum will separate all the molecules into two large groups: those close to the left minimum (A) and those close to the right one (B). Classification of the molecules that happen to be somewhere halfway between the two minima is not so unambiguous and depends on $s^{\ddagger}$, but, since there are so few of such poorly classifiable molecules, they will not significantly affect N_A or N_B.

Now let us simply ignore any molecules that happened to be in the state B at $t = 0$. That is, in Figure 5.1 we will be watching only the molecules represented by the dark circles. Those $N_A(0)$ molecules form a nonequilibrium ensemble prepared in state A; see Fig. 5.1. As the time evolves, some of them will cross over to state B (where $s > s^{\ddagger}$). We count the number $N_B(t)$ of those molecules. By fitting the time evolution of $N_B(t)$ and $N_A(t) = N - N_B(t)$ to Eq. 3.2 from Chapter 3, we can extract the rate coefficients for the transitions from A to B or B to A in much the same manner as an experimentalist would do it by preparing an initial population of the reactant A and monitoring its subsequent time evolution.

The second approach is a computational chemist's version of a single-molecule experiment. Instead of examining the collective behavior of many molecules, the trajectory $s(t)$ of just one molecule is monitored and its state (A or B, as defined above) is recorded, as a function of time. Provided the trajectory is long enough that a sufficient number of jumps between A and B have been observed, we then can compute the mean times the molecule dwells in A and B, which, as demonstrated in Chapter 3, are inversely proportional to the rate coefficients one seeks.

5.2 OVERCOMING THE RARE EVENT PROBLEM: TRANSITION STATE THEORY

Unfortunately, neither of the above two procedures is practical. The problem is that chemical reactions are rare events. More precisely, the difficulty arises from a enormous disparity of the timescales simultaneously present in the system. For example, atomistic-level description of the motion within the A state involves molecular vibrations, which, typically, occur on a picosecond timescale. This means that our computer code must output trajectories with better than a picosecond time resolution and, consequently, that it must be evaluating the molecular forces multiple times during a single picosecond so as to provide details of the motion during a single vibration. This is computationally expensive. Depending on the speed of our computer, the size of the system, and the level of description, it could easily take seconds to minutes to predict one picosecond of the molecule's life. On the other hand, if a large energy barrier separates A and B, it will take a huge number of molecular vibrations before a single transition from A and B can be observed. Indeed such a transition requires a large thermal fluctuation, whereby the molecule gains enough energy E to overcome the barrier, and, according to the Boltzmann distribution, the probability of such fluctuation should be proportional to $\exp(-E/k_B T)$. The latter probability can be very small if E significantly exceeds the thermal energy. Given the exponential sensitivity of this probability to the barrier height we thus expect the typical time t_A our molecule dwells in the state A to be practically any number that is longer than the vibrational timescale. That is, it can be a millisecond, or a second, or a year. Of course if this time is too long, say much longer than our lifespan, then we could decide we do not care about this particular transition because it does not affect our well-being and/or cannot be observed in a lab. But even if we are dealing with a relatively fast, from an everyday life's standpoint, process that occurs, say, at a millisecond timescale, and assuming a typical computation speed of 1 picosecond per a CPU second, it will take us 10^9 seconds ≈ 30 years to observe a single transition in the simulation. That is, our computer will spend 30 years generating information about details of molecular motion of the molecule A without ever telling us anything useful about the transition from A to B.

This rare event problem is perhaps one of the most important challenges in modern computational chemistry and biophysics. A book could easily be written on this subject alone. Described below are just some of the powerful tricks allowing one to circumvent this problem. Recall, from Chapter 3, that the quantity $k_{A \rightarrow B} t$, for sufficiently short times t, can be interpreted as the probability to make a transition from A to B, conditional upon initially being in A at $t = 0$. This probability can also be thought of as the conditional probability $T(B, t|A, 0)$ that the system is found in B at time t provided it was in A at $t = 0$.[1] We then have the following expression for

---

[1] Note that $T(B, t|A, 0)$ is, generally speaking, *not* equal to the probability of making a *single* transition from A to B within the time interval t because the state B can be attained at t via any sequence of multiple jumps that ends in B. At sufficiently small values of t, however, a single-jump trajectory is much more likely than one involving multiple jumps. If you are not satisfied with this argument, a more formal derivation of Eq. 5.2 is given in Section 4.5 of the preceding chapter.

the rate coefficient:

$$k_{A \to B} = \frac{d}{dt} T(B, t|A, 0)\bigg|_{t \to 0+} = w_A^{-1} \frac{d}{dt} J(B, t|A, 0)\bigg|_{t \to 0+}. \qquad (5.2)$$

Here $J(B, t|A, 0)$ is the joint probability to be in A at $t = 0$ and in B at time t and w_A is the equilibrium probability to be in A. It is convenient to express these probabilities in terms of correlation functions involving the following functions:

$$\theta_A(s) = \theta(s^\ddagger - s) = \begin{cases} 1, & \text{if } s < s^\ddagger \\ 0, & \text{if } s \geq s^\ddagger \end{cases} \qquad (5.3)$$

$$\theta_B(s) = 1 - \theta_A(s) = \theta(-s^\ddagger + s) = \begin{cases} 1, & \text{if } s \geq s^\ddagger \\ 0, & \text{if } s < s^\ddagger \end{cases}. \qquad (5.4)$$

Here $\theta(x)$ is the Heaviside step function (equal to one if $x > 0$ and zero otherwise) and, consequently, $\theta_{A(B)}(s)$ is one if the system is in state A(B) and zero otherwise. In terms of these functions, the equilibrium probability to be in A or B can be expressed by the average:

$$w_{A,B} = \langle \theta_{A,B} \rangle. \qquad (5.5)$$

Indeed, by evaluating this as a time average,

$$(1/\tau) \int_0^\tau dt \, \theta_{A(B)}[s(t)],$$

where $\tau \to \infty$, one simply gets the fraction of the time between 0 and τ spent in the state A(B).

Likewise, the joint probability of being in A at zero time and in B at time t can be expressed as the average of a product (i.e., a correlation function) of the form:

$$J(B, t|A, 0) = (1/\tau) \int_0^\tau dt' \theta_A[s(t')] \theta_B[s(t'+t)] = \langle \theta_A[s(0)] \theta_B[s(t)] \rangle. \qquad (5.6)$$

Here we have taken advantage of the time translation symmetry, which implies that only the time difference t between the two events (being in A and in B) is significant and so the time origin in an equilibrium correlation function can be shifted by an arbitrary amount t' without changing its value. Substitution of this into Eq. 5.2 gives

$$k_{A \to B} = \lim_{t \to 0+} k_{A \to B}(t), \qquad (5.7)$$

where we have defined the time-dependent quantity

$$k_{A \to B}(t) = \frac{\langle \theta_A(0) \dot{\theta}_B(t) \rangle}{w_A} = \frac{1}{w_A} \left\langle \frac{d\theta_A(-t)}{dt} \theta_B(0) \right\rangle$$

$$= -\frac{\langle \dot{\theta}_A(0) \theta_B(t) \rangle}{w_A} = \frac{\langle \dot{\theta}_B(0) \theta_B(t) \rangle}{w_A}. \qquad (5.8)$$

Here the shorthand

$$\theta_{A(B)}(t) \equiv \theta_{A(B)}[s(t)]$$

is used and the dot indicates the derivative with respect to time. In deriving Eq. 5.8, time translation symmetry was used again, which, in particular, implies that $\langle X(0)Y(t)\rangle = \langle X(-t)Y(0)\rangle$ for any X and Y that are functions of the system's coordinate $s(t)$. In addition, the identity $\dot{\theta}_A = -\dot{\theta}_B$, following from the definition of θ_A and θ_B, was used. Writing

$$\dot{\theta}_B(t) = d\theta_B[s(t)]/dt = d\theta[s(t) - s^{\ddagger}]/dt = \dot{s}(t)\delta[s(t) - s^{\ddagger}],$$

and recognizing that the derivative of the Heaviside step function $\theta(x)$ is Dirac's delta function $\delta(x)$–a fact easily established by integrating the delta function over x and using Eq. 4.5–we finally get

$$k_{A \to B}(t) = w_A^{-1}\langle \dot{s}(0)\delta[s(0) - s^{\ddagger}]\theta[s(t) - s^{\ddagger}]\rangle. \tag{5.9}$$

Compared to the two procedures outlined above, Eq. 5.9 offers tremendous computational advantage. Specifically, the delta-function appearing in this expression selects the trajectories that start with $s(0) = s^{\ddagger}$. That is, instead of launching trajectories from all over the place and patiently waiting for them to cross from one state to the other, we can focus on the ones caught in the act of already crossing the boundary between A and B at $t = 0$. Computationally, to evaluate Eq. 5.9, multiple trajectories are launched from $s^{\ddagger}$, each trajectory having an initial momentum drawn from the Maxwell-Boltzmann distribution. If a trajectory ends up in state B at time t then $\theta_B[s(t)] = 1$ so it contributes $\dot{s}(0)$ to the average; otherwise, its contribution is zero. If $s^{\ddagger}$ is selected near the top of the barrier, then we expect a large fraction of trajectories to yield nonzero contribution, rendering our procedure efficient.

So far we have not taken the limit $t \to 0+$ required by Eq. 5.7. At small values of t, however, we can approximate $s(t)$ by its Taylor expansion, $s(t) \approx s^{\ddagger} + \dot{s}(0)t$. Then $\theta[s(t) - s^{\ddagger}] = \theta[\dot{s}(0)t] = \theta[\dot{s}(0)]$ is nonzero only if the initial velocity $\dot{s}$ is positive (i.e., is directed from the reactant A to the product B state). Our procedure is thus further simplified by counting all the trajectories starting at $s^{\ddagger}$ with positive initial velocities and discarding the ones with negative velocities. In fact, it is easy to evaluate this short time limit of Eq. 5.9 analytically by performing a Boltzmann-weighted average (see Appendix B):

$$
k_{A \to B} = w_A^{-1} q^{-1} \iint\limits_{p>0} \frac{dp\,ds}{2\pi\hbar} \left(\frac{p}{m}\right) \delta(s - s^{\ddagger}) \exp\left(-\beta\left[\frac{p^2}{2m} + V(s)\right]\right)
$$

$$
= q_A^{-1} \exp\left[-\beta V(s^{\ddagger})\right] \int_0^\infty \frac{dp}{2\pi\hbar} \left(\frac{p}{m}\right) \exp\left(-\beta\frac{p^2}{2m}\right)
$$

$$
= \frac{k_B T}{2\pi\hbar q_A} \exp\left[-\beta V(s^{\ddagger})\right] \tag{5.10}
$$

where $p = m\dot{s}(0)$, q_A is the partition function of the system restricted to the state A (i.e., to the left of $s^{\ddagger}$), and $q = q_A + q_B$ is the total partition function.

Although Planck's constant appears in the above expression, it cancels out with the same constant in the expression for the reactant partition function q_A so that the final answer does not contain any signatures of quantum mechanics, as expected for a system that is being treated classically. For example, if the shape of the left well is well described by a parabola of the form

$$V(s) \approx V(s_A) + m\omega_A^2(s - s_A)^2/2$$

then the corresponding partition function is

$$
\begin{aligned}
q_A &\approx \frac{1}{2\pi\hbar} \int_{-\infty}^{\infty} dp \int_{-\infty}^{s^\ddagger} ds \exp\left[-\beta\left(\frac{p^2}{2m} + V(s_A) + m\frac{\omega_A^2(s - s_A)^2}{2}\right)\right] \\
&\approx \frac{k_B T}{\hbar\omega_A} \exp\left[\frac{-V(s_A)}{k_B T}\right],
\end{aligned}
\tag{5.11}
$$

which is obtained by replacing the upper integration limit in s by ∞, an approximation that assumes that the integrand vanishes away from the minimum before deviations from the harmonic approximation become appreciable. This result is nothing but the harmonic-oscillator partition function discussed in Appendix B. Substituting this into Eq. 5.10 we find

$$k_{A\to B} = \frac{\omega_A}{2\pi} \exp\left[-\frac{V(s^\ddagger) - V(s_A)}{k_B T}\right]. \tag{5.12}$$

The preexponential factor, equal to the inverse of the vibration period in the left well, can be thought of as the frequency, with which the system attempts to make the transition, while the exponential, equal to the Boltzmann probability of overcoming the energy barrier $V(s^\ddagger) - V(s_A)$, can be thought of as the probability of an attempt to be successful.

The rate coefficient for going backward can be determined in a similar manner,

$$k_{B\to A} = \frac{k_B T}{2\pi\hbar q_B} \exp[-\beta V(s^\ddagger)] \approx \frac{\omega_B}{2\pi} \exp\left[-\frac{V(s^\ddagger) - V(s_B)}{k_B T}\right]$$

where q_B is the partition function of the system in the right well approximated as that of a harmonic oscillator of frequency ω_B (cf. Appendix B). The rates thus estimated satisfy detailed balance,

$$k_{A\to B}/k_{B\to A} = q_B/q_A = w_B/w_A = \frac{\omega_A}{\omega_B} \exp\left[-\frac{V(s_B) - V(s_A)}{k_B T}\right]. \tag{5.13}$$

Eq. 5.10 is an example of the more general transition-state theory (TST), which will be derived in Section 5.7 for the more general case of many degrees of freedom. It is a very powerful result because it expresses a dynamical quantity, the reaction rate, in terms of equilibrium properties of the system. That is, evaluating Eq. 5.10 does not require running actual trajectories $s(t)$, as the end result is independent of the precise manner in which the system explores its energy landscape. The latter observation, however, is rather perplexing: Consider, for instance, a molecule that is immersed in a high viscosity solvent. TST predicts that the rate of any chemical reaction would remain independent of how viscous the solvent is, as long as the solvent viscosity does not affect the underlying potential $V(s)$ describing this reaction. This does not make sense: Increasing the solvent viscosity is naturally expected to slow the reaction down. Another, even more evident, problem exists with Eq. 5.10: This result is strongly dependent on the choice of $s^{\ddagger}$, the boundary separating the reactants from the product. Although it appears most sensible to place this boundary at the top of the barrier, $s^{\ddagger} = s_m$ (Fig. 5.1), nowhere in our derivation did we assume this to be generally the case. The strong, exponential dependence of the TST rate on the definition of the states A and B is inconsistent with the experimental observation that, in the case of first-order kinetics, the rate coefficients are robust and independent of the precise manner in which they are measured.

5.3 WHY TRANSITION STATE THEORY IS NOT EXACT

To resolve the above difficulties, it helps to examine the physical picture suggested by our expression for the rate, Eq. 5.9, in more detail. This equation prescribes us to launch trajectories at $t = 0$ from $s = s^{\ddagger}$. The flux of those trajectories, i.e., the number of trajectories crossing the boundary between A and B per unit time, is proportional to $\dot{s}(0)$. However they are "reactive," i.e., contribute to the rate, only if they end up on the product side. This condition is enforced by the step function $\theta[s(t) - s^{\ddagger}]$, which excludes any trajectories that are to the left of the boundary at time $t > 0$ after being launched. Previously, we have claimed that the trajectories that have positive velocities at $t = 0$ are all reactive. Although formally this appears to be correct because of the $t \to 0+$ limit implied in Eq. 5.7, this claim seems unphysical. Suppose, for example, that the boundary separating A from B is placed to the left of the barrier top, $s^{\ddagger} < s_m$, as in Fig. 5.1. The trajectories that cross the boundary at $t = 0$ with a positive velocity do not necessarily have enough energy to surmount the rest of the barrier and so they will recross back to the reactant. It makes sense not to count such trajectories when calculating the rate. Indeed, if we assume energy conservation along the trajectories and replace the lower boundary for the momentum integration in Eq. 5.10 by the minimum momentum p_m required to surmount the remaining barrier $V(s_m) - V(s^{\ddagger})$,

$$p_m = \sqrt{2m[V(s_m) - V(s^{\ddagger})]},$$

then we get a different rate estimate,

$$k_{A \to B} = q_A^{-1} \exp[-\beta V(s^{\ddagger})] \int_{p_m}^{\infty} \frac{dp}{2\pi\hbar} \left(\frac{p}{m}\right) \exp\left(-\beta \frac{p^2}{2m}\right) = \frac{k_B T}{2\pi\hbar q_A} \exp[-\beta V(s_m)],$$

(5.14)

which appears a lot more plausible than Eq. 5.10 as it is independent of $s^{\ddagger}$ and is determined by the energy required to reach the top of the barrier. Nevertheless, it still remains to explain, formally, why and how Eq. 5.10 fails.

Let us reexamine the starting point of our derivation, Eq. 5.2. This equation assumes a system that undergoes instantaneous jumps among discrete states. In reality, of course, such jumps are not truly instantaneous. Because of a finite jump duration, we have to be more careful as to what we call a transition. Consider, for example, the trajectory $s(t)$ of a particle in a symmetric double-well potential (Fig.5.2). I have generated this trajectory by numerically solving a Langevin equation of motion,

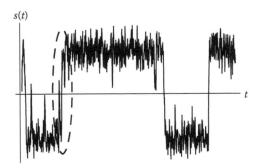

$s(t)$

t

FIGURE 5.2 Simulated trajectory of a molecular system modeled as a particle undergoing Langevin dynamics in a symmetric double-well, with a single transition highlighted.

as discussed in Chapter 4. During the portion of the trajectory highlighted in the Figure 5.2, it exhibits a marked transition from one well to the other. But how many transitions do we actually see? If we formally call a transition every event the midpoint between the two wells is crossed, we can easily count several such events. On the other hand, the physical picture introduced in the preceding chapter implies that we should count them as only one real transition. Indeed, a true transition involves achieving local thermal equilibrium within the reactant or the product well every time a transition to this well is made. Clearly, no such equilibrium is reached while the trajectory makes several quick excursions across the midpoint without even reaching the vicinity of either potential minimum. And so such excursions should not be viewed as true transitions.

Experimental considerations lend further support to this argument. Indeed, quick recrossings of $s^{\ddagger}$ often take place at pico- to nanosecond timescales comparable to those of local molecular motions. The time resolution of single-molecule measurements is usually insufficient to observe such events. An experimentalist therefore sees a smoothed version of Figure 5.2, in which any fast recrossing events are washed out.

These considerations suggest that it is necessary to modify Eq. 5.7 to avoid counting fast recrossings of the boundary between A and B as true transitions. The correct rate should be the value of $k_{A \to B}(t)$ not at $t \to 0+$ but at some intermediate time t that is longer than the duration τ_{tr} of a typical molecular transition. On the other hand t must be shorter than the typical time the molecule dwells in the reactant and the product state, i.e., $t \ll k_{A \to B}^{-1}, k_{B \to A}^{-1}$. This condition ensures that the time t, while nonzero microscopically, remains, effectively, to be the zero-time limit as far as the dynamics of the coarse-grained two-state model is concerned. Obviously, the above two conditions are compatible only if the typical duration of an individual transition event τ_{tr} is much shorter than the typical dwell times in the reactant and the product states.

Our proposal is thus to replace Eqs. 5.2 and 5.7 by

$$k_{A \to B} = \frac{d}{dt} T(B, t | A, 0) \Big|_{\tau_{tr} \ll t \ll k_{A \to B}^{-1}, k_{B \to A}^{-1}}$$

$$= w_A^{-1} \langle \dot{s}(0) \delta[s(0) - s^{\ddagger}] \theta[s(t) - s^{\ddagger}] \rangle \big|_{\tau_{tr} \ll t \ll k_{A \to B}^{-1}, k_{B \to A}^{-1}}. \qquad (5.15)$$

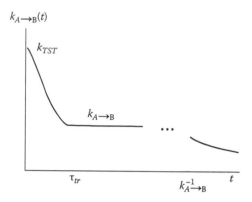

FIGURE 5.3 Typical time dependence of $k_{A \to B}(t)$ exhibits a plateau at intermediate times, which corresponds to the true rate. Zero-time limit of $k_{A \to B}(t)$ yields a higher value corresponding to the transition-state theory rate. At long times, $k_{A \to B}(t)$ becomes time-dependent again, as prescribed by the master-equation kinetics.

Naturally, Eq. 5.15 is meaningful only if $k_{A \to B}(t)$ is insensitive to the precise value of time t. Indeed, for systems amenable to a first-order kinetic description a plateau in the function $k_{A \to B}(t)$, corresponding to the true rate coefficient, is commonly observed (Fig. 5.3).

5.4   THE TRANSMISSION FACTOR

The rate coefficient computed from Eq. 5.15 is independent of the precise choice of the boundary $s^{\ddagger}$ separating the reactant A and the product B. To demonstrate this, consider a single transition event shown in Fig. 5.4. During this event, the molecule, having arrived from the reactant state, crosses the boundary $s = s^{\ddagger}$ five times before finally committing to the product state. We, however, do not regard the rapid recrossings of $s^{\ddagger}$ as genuine transitions and so, rather than five transitions, only one is actually observed in Fig. 5.4. To estimate the transition rate we launch multiple trajectories from $s^{\ddagger}$ and inquire into their fate after a time t, as prescribed by Eq. 5.15. This time is long enough that each trajectory has committed itself either to B or to A. If we could sample every possible trajectory in this manner, then we would generate 5 different fragments of the same trajectory representing the event with 5 crossings. For example, the fragment that starts at the second crossing is shown in Fig. 5.4 on the right. Because each such trajectory fragment enters into the rate expression with a factor $\dot{s}$ equal to the velocity at the crossing, the first crossing contributes a positive flux, the second crossing contributes a negative flux and so forth. Thus we have 3 positive fluxes and 2 negative ones. The contributions from the even and the odd crossings happen to have the same absolute values but opposite signs.[2] Four out

---

[2] This is a consequence of the Liouville theorem (see, e.g., [1]), which necessitates that, for a conservative system, fluxes from swarms of trajectories in the vicinity of any given trajectory are conserved. This remains true even for non-conservative systems such as those described by Langevin dynamics because

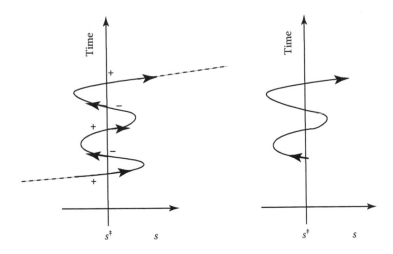

FIGURE 5.4 A trajectory that crosses the dividing point $s^{\ddagger}$ five times (left) will contribute into Eq. 5.15 five times. One of such contributions, a trajectory fragment that contributes a negative flux, is shown on the right.

of five contributions will therefore cancel out and we are left with a single flux for each transition event. Moreover, the value of this flux will remain the same regardless of the value $s^{\ddagger}$ since, although the total number of crossings may change (e.g., can be 1 or 3 in the example shown in Fig. 5.4), it will always remain an odd number, resulting in the same positive flux because of the cancellation of all the fluxes but one. Therefore we have shown that, indeed, the rate $k_{A \to B}$ estimated from Eq. 5.15 remains independent of $s^{\ddagger}$.

Because each trajectory effectively contributes into Eq. 5.15 only once, we can modify this equation such that only one crossing, say the last one, is counted. That is, we write

$$k_{A \to B} = w_A^{-1} \langle \dot{s}(0) \delta[s(0) - s^{\ddagger}] \chi[\dot{s}(0)] \rangle, \qquad (5.16)$$

where $\chi(v) = 1$ if (i) the trajectory $s(t)$ that starts from $s^{\ddagger}$ with the velocity $\dot{s}(0) = v$ goes straight to the product B without ever recrossing $s^{\ddagger}$ and (ii) the trajectory $s(-t)$, which starts from $s^{\ddagger}$ with the velocity $-v$, ends up in the reactant A (whether or not it recrosses does not matter). We set $\chi(v) = 0$ if either of the above two conditions is not satisfied. Each full reactive trajectory, obtained by combining $s(t)$ with the time-reversed $s(-t)$, is then counted by Eq. 5.16 exactly once.

By construction, the function $\chi(v)$ is zero if $v < 0$. If one replaces it by the Heaviside step function, $\theta(v)$, then the transition-state theory rate coefficient is immediately recovered. The function $\chi(v)$ is, however, more restrictive than $\theta(v)$ as the positivity of the initial velocity does not guarantee that either of the two above conditions is satisfied. Therefore we conclude that the true rate coefficient $k_{A \to B}$ cannot be greater

---

the additional nonconservative terms (frictional forces and random noise) can be viewed as arising from interaction with surrounding molecules and so the extended system including the surrounding molecules is still conservative.

than the transition-state theory one. The transition-state theory, therefore, always provides an upper bound on the true rate. This suggests that the TST rate estimate can be improved via a variational approach, whereby the TST rate coefficient is minimized with respect to the location of the boundary $s^{\ddagger}$. For example, the best choice for $s^{\ddagger}$ in a simple potential shown in Fig. 5.1 would obviously be the top of the barrier, where the Boltzmann factor $\exp[-\beta V(s^{\ddagger})]$ attains its minimum. If one is lucky, such optimal choice of boundary will yield the exact transition rate. If so, $s^{\ddagger}$ can be interpreted as the true transition state or the point of no return: Any trajectory crossing $s^{\ddagger}$ in the direction of the product B is committed to accomplish the transition without returning, even for a fleeting moment, to A.

Even with the best choice of $s^{\ddagger}$, however, the TST rate is not guaranteed to be exact. The ratio of the true rate and the TST rate,

$$\kappa_{A \to B} = k_{A \to B} / k_{A \to B}^{TST} = \frac{\langle \dot{s}(0) \delta[s(0) - s^{\ddagger}] \chi[\dot{s}(0)] \rangle}{\langle \dot{s}(0) \delta[s(0) - s^{\ddagger}] \theta[\dot{s}(0)] \rangle}, \qquad (5.17)$$

is commonly referred to as the transmission factor. This factor is always less than or equal to one. It can also be interpreted as a Maxwellian-flux-weighted probability that a trajectory launched from $s^{\ddagger}$ satisfies the requirements to have a nonzero $\chi[\dot{s}(0)]$, or, equivalently, belongs to a reactive trajectory that has crossed the boundary $s^{\ddagger}$ for the last time before committing to the product.

The transmission factor has a further important property: It is the same for both the direct and the reverse reactions, allowing us to drop the subscript indicating the direction of the transition:

$$\kappa_{A \to B} = \kappa_{B \to A} = \kappa. \qquad (5.18)$$

Indeed, consider the detailed balance requirement:

$$\frac{k_{A \to B}}{k_{B \to A}} = \frac{k_{A \to B}^{TST} \kappa_{A \to B}}{k_{B \to A}^{TST} \kappa_{B \to A}} = \frac{q_B}{q_A}.$$

In view of the fact that the TST rate coefficients already satisfy the detailed balance condition (cf. Eq. 5.13), we immediately obtain Eq. 5.18. The identity of the forward and backward transmission factors is a direct consequence of time reversibility, which will be discussed in greater detail in the next chapter. Specifically, time reversibility implies that the ensemble of "forward" trajectories $s(t)$ is indistinguishable from that of backward ones, $s(-t)$. Since the time reversal, $t \to -t$, transforms any A-to-B reactive trajectory into a B-to-A reactive trajectory, it immediately follows that the statistics of the reactive trajectories is exactly the same for both directions, leading to the same value of the transmission factor.

5.5 RELATIONSHIP BETWEEN THE TRANSMISSION FACTOR AND THE NUMBER OF CROSSINGS

From the discussion above, it is clear that transition-state theory would be exact if trajectories never recrossed the boundary between A and B. It then makes sense that the transmission factor should be related to the average number of crossings involved in a single transition event. Consider, for example, a simple probabilistic model, in which recrossings are statistically independent. That is, if a trajectory is launched from $s^{\ddagger}$, it has a fixed probability w to commit to the reactants or products (depending on the direction it was launched) without ever coming back. It will therefore return to $s^{\ddagger}$ with a probability equal to $1 - w$. Assuming w is independent of the velocity $\dot{s}$ at the crossing, we can use Eq. 5.17 to estimate κ as the probability w of committing to the product times the probability of making an even number of recrossings upon being launched (backwards) from $s^{\ddagger}$. The probability of launching a trajectory that will make exactly n recrossings before committing to A or B is given by $w(1 - w)^n$. Therefore, the probability of making an even number of recrossings is

$$w + w(1 - w)^2 + w(1 - w)^4 + \cdots = \frac{w}{1 - (1 - w)^2}$$

and so

$$\kappa = \frac{w^2}{1 - (1 - w)^2}.$$

If $w \ll 1$ then this expression is further simplified to give

$$\kappa \approx w/2.$$

On the other hand, w can be related to the average number of crossings made by any trajectory that has reached $s^{\ddagger}$. Indeed, if we count the crossings starting from the first one, the probability for exactly one crossing is equal to the probability w of leaving $s^{\ddagger}$ without recrossing. Likewise, the probability of making n crossings is

$$w_n = w(1 - w)^{n-1} \tag{5.19}$$

and the average number of crossings is

$$\langle n \rangle = \sum_{n=1}^{\infty} n w(1 - w)^{n-1} = -w(d/dw) \sum_{n=1}^{\infty} (1 - w)^n = 1/w.$$

We conclude that, in the limit $\langle n \rangle \gg 1$, the transmission factor is inversely proportional to the average number of crossings:

$$\kappa \approx \frac{1}{2\langle n \rangle}.$$

Of course, there is no real reason to believe that the simple, velocity-independent crossings statistics of Eq. 5.19 is strictly satisfied for any molecular system.

5.6 THE TRANSMISSION FACTOR FOR LANGEVIN DYNAMICS

Earlier in this book, the Langevin equation was introduced as a simple physical model to describe the behavior of a molecule in a condensed phase environment. This equation accounts for the forces caused by the surrounding molecules by introducing a random force $R(t)$ and a friction force proportional to the velocity:

$$m\ddot{s} = -V'(s) - \eta\dot{s} + R(t), \tag{5.20}$$

Here η is a friction coefficient, which is related to the diffusion coefficient D through the Einstein-Smoluchowski relationship,

$$D = \frac{k_B T}{\eta}.$$

The random force further satisfies the fluctuation-dissipation relationship

$$\langle R(t)R(t') \rangle = 2\eta k_B T \delta(t - t').$$

The dynamics predicted by Eq. 5.20 is stochastic. That is, even if a trajectory is launched from the top of the barrier ($s^{\ddagger} = s_m$) towards the product B (Fig. 5.1), the random force may cause it to reverse its direction, possibly multiple times, resulting in a trajectory of the type shown in Figure 5.2 (in fact, Fig.5.2 was generated using a Langevin equation). It is then clear that transition state theory is not adequate for this case unless supplemented with an appropriate transmission factor.

As discussed in Section 5.3, the transmission factor κ can be calculated numerically by repeatedly launching trajectories from the top of the barrier and counting the reactive ones. Initiated at the top, a trajectory has nowhere to go but downhill. As long as it remains in close proximity to the barrier top, the trajectory may keep on recrossing its starting point, s_m, multiple times, but once it travels far enough that its energy becomes lower than that of the barrier top by, say, a few $k_B T$, the return to the starting point becomes unlikely. The trajectory, therefore, becomes committed to either the reactant or the product side of the barrier and, for the purpose of computing κ, its further evolution is immaterial. Thus, in a practical implementation of such a calculation one can stop following each trajectory once the distance from the top, $|s - s_m|$, exceeds a certain value. If, within this distance range, the potential is well approximated by its second-order Taylor expansion,

$$V(s) \approx V(s_m) + V''(s_m)(s - s_m)^2/2,$$

then, for the purpose of calculating the transmission coefficient, the actual potential $V(s)$ can be replaced by a parabolic barrier

$$V(s) = V(s_m) - m(\omega^{\ddagger})^2(s - s_m)^2/2,$$

where

$$\omega^{\ddagger} = \sqrt{-V''(s_m)/m}$$

is the upside down frequency at the top of the barrier.

Introducing dimensionless time and length variables,

$$\theta = \frac{m(\omega^{\ddagger})^2}{\eta} t$$

$$\xi = (s - s_m)\sqrt{\frac{m(\omega^{\ddagger})^2}{k_B T}},$$

we can rewrite Eq. 5.20 in a dimensionless form,

$$(m\omega^{\ddagger}/\eta)^2 d^2\xi/d\theta^2 = -d\xi/d\theta + \xi + \rho(\theta), \tag{5.21}$$

where the dimensionless force ρ satisfies the fluctuation-dissipation relation

$$\langle \rho(\theta)\rho(\theta') \rangle = 2\delta(\theta - \theta').$$

Now comes an essential point: the dynamics described by Eq. 5.21 depends on the value of a single dimensionless parameter,

$$m\omega^{\ddagger}/\eta, \tag{5.22}$$

and nothing else. Since the transmission factor can be computed from Eq. 5.21, *its value must be entirely determined by this dimensionless parameter*, i.e., we have

$$\kappa = \psi(m\omega^{\ddagger}/\eta), \tag{5.23}$$

where $\psi(x)$ is some function. The above argument exemplifies the idea of *dimensional analysis*: Certain essential relationships between physical quantities can be established by merely analyzing the units in which they are measured. Here, we know that the transmission factor is a dimensionless quantity and, as such, can only depend on a dimensionless combination of the mass m, friction coefficient η, and barrier frequency $\omega^{\ddagger}$. Any such dimensionless combination happens to be either the ratio of Eq. 5.22 or some function ψ of this ratio, thus leading to Eq. 5.23.

Dimensional analysis can only get us so far. It cannot generally give us the precise form of the function ψ appearing in Eq. 5.23. It turns out, however, that this function can be determined in the overdamped, high-friction limit (see Chapter 4). Let us turn back to the original Langevin equation, Eq. 5.20. As explained in Chapter 4, in the overdamped limit the second derivative term can be neglected, which gives

$$0 = -V'(s) - \eta ds/dt + R(t),$$

or, equivalently,

$$\frac{ds}{d(t/\eta)} = -V'(s) + R(t),$$

where the fluctuation-dissipation relationship for the force R can further be rewritten in the form[3]

$$\langle R(t)R(t') \rangle = 2k_B T \delta \left(\frac{t}{\eta} - \frac{t'}{\eta} \right).$$

Note that the time and the friction coefficient enter in the above two formulas through their ratio, t/η. For example doubling both the time and the friction coefficient results in no change in the formulas describing the dynamics. This means that *the timescale of any process as predicted by an overdamped Langevin equation must be proportional to the friction coefficient η*. This includes, in particular, the average time spent in the reactant state A before making a transition to B. Accordingly, the inverse of this time, equal to $k_{A \to B}$, must be inversely proportional to η. But since the transition-state rate does not depend on the friction coefficient, the transmission coefficient κ must itself

---

[3] Here we are using the identity $\delta(bt) = \delta(t)/b$, where b is a real number that is not equal to zero. This identity can be easily verified by substituting it into Eq. 4.5.

be inversely proportional to η. Combined with Eq. 5.23, this conclusion necessitates that

$$\kappa = c \frac{m\omega^{\ddagger}}{\eta}, \tag{5.24}$$

where c is some numerical constant. The reader should pause and observe here that we have practically solved (to within a numerical constant) the problem of finding the transmission factor in the overdamped case without ever solving the equation that governs the system's dynamics!

A more general result was obtained in 1940 by Hendrik Kramers, a Dutch physicist who has made important contributions into many areas of physics, particularly quantum mechanics. Kramers' result [2],

$$\kappa = \sqrt{1 + \left(\frac{\eta}{2m\omega^{\ddagger}}\right)^2} - \frac{\eta}{2m\omega^{\ddagger}},$$

is in agreement with Eq. 5.23. In the overdamped limit, where we have $\frac{\eta}{m\omega^{\ddagger}} \gg 1$, this gives Eq. 5.24 with a factor c that conveniently turns out to be equal to 1. Finally, multiplying the TST rate, Eq. 5.12, by κ, we obtain the following expression for the transition rate coefficient in the overdamped case:

$$\begin{aligned} k_{A \to B} &= \frac{m\omega_A \omega^{\ddagger}}{2\pi\eta} \exp\left[-\frac{V(s_m) - V(s_A)}{k_B T}\right] \\ &= \frac{\sqrt{V''(s_A)|V''(s_m)|}}{2\pi\eta} \exp\left[-\frac{V(s_m) - V(s_A)}{k_B T}\right]. \end{aligned}$$

This rate coefficient is inversely proportional to the friction coefficient and thus to the solvent viscosity, a trend anticipated above.

5.7 EXTENSION TO MANY DEGREES OF FREEDOM

Our one-dimensional model of reaction dynamics is generally inadequate for molecules, which have many degrees of freedom. Fortunately, its extension to higher-dimensional systems is fairly straightforward. To see this, let us slightly change our perspective and think of s as some collective variable that describes the progress of the reaction from A to B. That is, while a complete description of the molecule includes the coordinates of all of its N atoms, i.e., the $3N$-dimensional vector

$$\mathbf{R} = (x_1, y_1, z_1, x_2, y_2, z_2, \ldots, x_N, y_N, z_N),$$

the state A can be distinguished from B by measuring the value of the scalar function $s = s(\mathbf{R})$. We will call s a "reaction coordinate" describing the reaction. The existence of a suitable reaction coordinate for any experimentally observable chemical reaction is virtually guaranteed by the very fact that the reaction can be observed. Indeed, most experimental measurements of chemical kinetics rely on the time dependence of a single quantity (aka experimental signal). Thus the most natural, though not necessarily most convenient, choice for s would be the experimental signal itself!

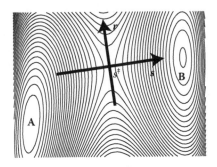

FIGURE 5.5 Harmonic transition-state theory in two dimensions. At the saddle point, the molecule's potential attains a maximum along the reaction coordinates while it corresponds to a minimum along the transverse coordinate r.

When s is a suitable reaction coordinate, Eq. 5.15 remains true regardless of the underlying dimensionality of the system. In practice, an intermediate value $s^{\ddagger}$ must be specified to distinguish between A and B. The condition $s(\mathbf{R}) = s^{\ddagger}$ defines a $3N - 1$-dimensional hypersurface in the system's $3N$-dimensional space. This hypersurface divides the space into the reactant (A) and product (B) regions. To compute $k_{A \to B}$, we launch, at $t = 0$, trajectories each starting from this hypersurface. In terms of the reaction coordinate s, the initial velocity is given by $\dot{s} = \nabla s \dot{\mathbf{R}}$, where $\dot{\mathbf{R}}$ is the system's velocity vector chosen from the Maxwell-Boltzmann distribution and ∇ stands for the gradient. Likewise, the initial positions on the hyperplane must be chosen according to the Boltzmann statistics. Describing exactly how to do this may take us a bit too far from the topic of this book—again, the reader interested in such details should consult more specialized literature [3]. Instead, I will now develop an approximation that generalizes the one-dimensional transition-state theory introduced above.

Recall that TST essentially states that the rate of a transition equals the Boltzmann probability of reaching the top of the barrier times an "attempt frequency" prefactor. In one dimension, the top of the barrier simply means the energy difference between the potential maximum (the optimal transition state) and the minimum corresponding to the reactants. What would be the corresponding barrier in more than one dimension? Say, if we pretend that two degrees of freedom are sufficient to describe the reaction of interest, then the molecule's potential, as a function of those degrees of freedom, may look like the "topographic map" shown in Figure 5.5. As in one dimension, it has two minima corresponding to the reactant and the product states. Think of a mountainous landscape. A minimum is a mountain valley or the bottom of a lake. To escape from this minimum to another valley, it is necessary to cross a mountain ridge. The easiest way to cross such a ridge is to head over a mountain pass, i.e., the lowest point on the ridge. A path crossing the mountain pass is optimal from a mountain climber's point of view because it requires the least energy expense. Because the likelihood of encountering a molecular state with a potential energy V is proportional to the Boltzmann factor, $\exp(-\beta V)$, a typical molecular transition should, likewise, follow a pathway that is not too far from the one minimizing the potential energy barrier encountered in the transition. We therefore expect it to cross the barrier in the vicinity of the "mountain pass." Mathematically, a mountain pass is a saddle point. Similarly

to a maximum or a minimum, it is a point where the gradient of the potential energy
vanishes. It is, however, neither a minimum nor a maximum; instead, it appears to be
a maximum when moving in one direction (e.g., from the reactants to the products or
from one mountain valley to the other) but is a minimum if one tries to walk along the
mountain ridge. It is then possible to introduce two coordinates, the unstable direction
s, along which one sees a barrier, and the stable direction r, along which there one
sees a minimum, such that the energy of the system in the vicinity of the saddle point
is approximated by its second order expansion:

$$V(s, r) \approx V^{\ddagger} + \frac{m(\omega_r^{\ddagger})^2}{2} r^2 - \frac{m(\omega_s^{\ddagger})^2 (s - s^{\ddagger})^2}{2}. \tag{5.25}$$

Here, $V^{\ddagger}$ is the energy at the saddle point located at $s = s^{\ddagger}$, $r = 0$. The linear terms
in Eq. 5.25 are absent because of the above condition that the gradient of $V(s, r)$
must vanish at the saddle and the cross-terms proportional to the product $(s - s^{\ddagger})r$
are eliminated by choosing appropriate directions of s and r.

It is natural to use the unstable mode s as the reaction coordinate. It is a very
convenient choice because, as long as the harmonic expansion of Eq. 5.25 holds,
the motion along r (which we are not particularly interested in) is decoupled from
the motion along s. For the motion along s, we then can develop a transition-state
approximation in exactly the same manner as we did in one dimension. That is, we
replace $\theta(s(t) - s^{\ddagger})$ by $\theta(\dot{s}(0))$ in Eq. 5.9 to obtain

$$k_{A \to B} = \frac{q}{q_A} \langle \dot{s}(0) \delta[s(0) - s^{\ddagger}] \theta[\dot{s}(0)] \rangle$$

$$= \frac{q}{q_A} \frac{1}{q} \int_{p_s > 0} \frac{dp_s ds dp_r dr}{(2\pi\hbar)^2} \frac{p_s}{m} \delta(s - s^{\ddagger}) \exp\left[-\beta \frac{p_s^2 + p_r^2}{2m} - \beta V(s, r)\right]$$

$$= \frac{q_r}{q_A} \frac{k_B T}{2\pi\hbar} e^{-\beta V^{\ddagger}}, \tag{5.26}$$

where p_s and p_r are the momenta corresponding to s and r, respectively, and

$$q_r = \frac{k_B T}{\hbar \omega_r^{\ddagger}}$$

is the harmonic-oscillator partition function for the transverse degree of freedom
(cf. Appendix B). When the molecule's potential energy surface depends on $n > 2$
variables,[4] we proceed in the same manner. We look for the "mountain pass" on the
n-dimensional potential energy landscape. This is a saddle point, at which we have
one unstable direction s and $n - 1$ stable directions such that the potential energy can
be approximated by

$$V(s, r_1, r_2, \ldots, r_{n-1}) \approx V^{\ddagger} + \sum_{i=1}^{n-1} \frac{m(\omega_i^{\ddagger})^2}{2} r_i^2 - \frac{m(\omega_s^{\ddagger})^2 (s - s^{\ddagger})^2}{2}, \tag{5.27}$$

---

[4] Although the total number of degrees of freedom in a molecule consisting of N atoms is $3N$, the number
n of relevant degrees of freedom is lower because the translation or rotation of the molecule as a whole
does not change its potential energy.

leading to the following rate estimate:

$$k_{A \to B} = \frac{k_B T}{2\pi\hbar} \frac{\prod_{i=1}^{n-1} q_i}{q_A} e^{-\beta V^{\ddagger}}, \tag{5.28}$$

where $q_i = k_B T/(\hbar\omega_i^{\ddagger})$.

Eq. 5.28 can also be rewritten as:

$$k_{A \to B} = \frac{k_B T}{2\pi\hbar} \frac{q^{\ddagger}}{q_A} = \frac{k_B T}{2\pi\hbar} e^{-\beta \Delta G^{\ddagger}}. \tag{5.29}$$

Here $q^{\ddagger}$ is the partition function of the "activated complex," defined as the system confined to the transition-state hyperplane, $s = s^{\ddagger}$. That is, this partition function is evaluated by setting $s = s^{\ddagger}$ and dropping integration over the momentum p_s conjugate to s. The free energy cost of forming such a complex is the activation free energy

$$\Delta G^{\ddagger} = -k_B T \ln(q^{\ddagger}/q_A).$$

Written in the form of Eq. 5.29 is how most chemists know transition-state theory. It was originally derived in 1935 by Henry Eyring, and, independently, by M.G. Evans and Michael Polanyi. Their work has provided a theoretical justification of the empirical relationship established by Arrhenius in the late 19th century, according to which the temperature dependence of most reaction rates has a temperature dependence of the form $k = A \exp(-E/k_B T)$. The original derivation of this result treated the activated complex as a real chemical species. As a result, the TST equation was obtained through a (brilliant) heuristic rather than rigorous mathematics and, consequently, initially received with considerable skepticism. In fact, the *Journal of Chemical Physics* initially rejected Eyring's paper, although the decision was later reversed. The derivation given here takes advantage of the developments that took place in chemical physics many decades after the Eyring-Evans-Polanyi theory; it thus avoids any ad hoc assumptions and clarifies the mathematical approximations involved.

The preexponential factor appearing in Eq. 5.29 has units of inverse time and, at room temperature, is equal to

$$\nu = \frac{k_B T}{2\pi\hbar} \approx 6 \times 10^{12} s^{-1}.$$

Given its independence of the molecular properties, this quantity is often treated as a fundamental time constant that sets the timescales of all reaction rates. Moreover, it is sometimes viewed as setting the "speed limit" of chemical reactions. That is, in the hypothetic case of a vanishing free energy barrier, $\Delta G^{\ddagger} \to 0$, the rate appears to achieve its maximum possible value equal to ν. These considerations are, however, fundamentally flawed. It is easy to see that they cannot be right simply because—as already noted—the rate expression for a system that obeys classical mechanics must not contain Planck's constant, as the latter appears nowhere in Newtonian mechanics. Eq. 5.29, contrary to its appearance, does not really contain Planck's constant because the $\hbar$ that you see in the prefactor cancels out with the one buried in the ratio of the two partition functions, $q^{\ddagger}/q_A$. Indeed, the constrained partition function of the activated complex, $q^{\ddagger}$, has fewer integrations over coordinates and momenta than does q_A and so the ratio of the two is proportional to $\hbar$. However the timescale set by ν explicitly contains an $\hbar$ that does not cancel with anything and so this timescale cannot have any physical significance for a classical rate process. Indeed, we saw that, in the one-dimensional case, the physically correct prefactor comes out to be the oscillation frequency in the reactant well (cf. Eq. 5.12).

If ν is not a physically meaningful preexponetial factor then what is? To answer this question, let us further estimate the reactant's partition function, q_A, using a harmonic approximation. The approximation provided by Eq. 5.27 is clearly a bad one as far as the calculation of q_A is concerned because it does not even predict the existence of the reactant potential minimum. Instead, we can use a similar local expansion in the vicinity of the minimum:

$$V(x_1, x_2, \ldots, x_n) \approx V_A + \sum_{i=1}^{n} \frac{m\omega^2}{2} x_i^2, \qquad (5.30)$$

where V_A is the energy at the minimum, x_i represent the normal modes of the molecule, and ω_i are the corresponding frequencies. Unlike the case of the transition state, there is no unstable mode in this case. In this approximation, the partition function q_A is simply the product of the harmonic-oscillator partition functions,

$$q_A = e^{-\beta V_A} \prod_{i=1}^{n} \frac{k_B T}{\hbar \omega_i}.$$

Substituting this result into Eq. 5.28, we find:

$$k_{A \to B} = \frac{1}{2\pi} \frac{\prod_{i=1}^{n} \omega_i}{\prod_{i=1}^{n-1} \omega_i^{\ddagger}} e^{-\beta(V^{\ddagger} - V_A)}.$$

As expected, the new prefactor does not contain Planck's constant. It is also easy to see that Eq. 5.12 is recovered in one dimension.

5.8 REACTION KINETICS IN COMPLEX SYSTEMS: FLOPPY CHAIN MOLECULES, RANDOM WALKS, AND DIFFUSION CONTROLLED PROCESSES

The theory of reaction rates described so far assumed the existence of a single potential energy minimum that could be identified with the reactant state and of a single saddle (or a maximum in 1D) corresponding to the transition state. This assumption is often satisfied when the reaction in question involves sufficiently small molecules. It is however often invalid for conformational rearrangements occurring in larger systems such as, e.g., the polymeric molecules essential for life. Protein folding, in particular, was mentioned in Chapter 3 (Fig.3.1) as an example of a process that is often described by two-state, first-order kinetics. Yet its reactant state does not correspond to any well-defined protein structure. Rather, it includes diverse arrangements of the molecular

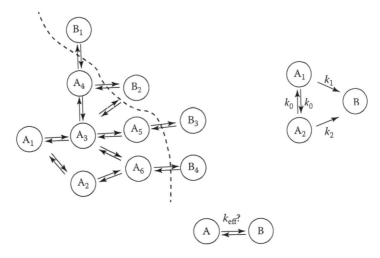

FIGURE 5.6 When studying chemical transformations involving large polyatomic molecules one often treats a collection of distinct states (corresponding, e.g., to different potential energy minima) as a single state. The unfolded state of a protein, for example, includes multiple conformations of the same molecule. Under what conditions can we approximate such a complex kinetic network as a simple two-state process obeying first-order kinetics, where some of the states are lumped into a combined reactant state (A) and the rest into a product state (B), as shown at the bottom? Shown on the right is a simple 3-state model studied in detail in the text.

chain, the number of which is, in fact, so large (see Section 4.6) that only their vanishingly small subset will be encountered over the lifetime of the organism that had produced the protein, or even over the lifetime of the Universe. Why can a two-state model be applied to describe such a system?

Consider a network of interconnected states such as the one shown in Fig.5.6, left. Each state may correspond to an individual PES minimum and first-order rate description is applicable to the transitions between individual states. However the observable "reactant" and "product" are now collective states that consist of reactant substates (A_n, $n = 1, 2, \ldots$) and the product substates B_n. Let $w_{A_n \to B}(t)$ be the distribution of the time t it takes to reach, for the first time, the collective state B having started from A_n. Then the probability distribution $w_{A \to B}(t)$ to reach B from any state in A can be calculated simply as the average of $w_{A_n \to B}(t)$ weighted by the probabilities of occupying each initial substate. On the other hand, the validity of first-order kinetics as applied to the collective states A and B requires the exponential distribution: $w_{A \to B}(t) = k_{A \to B} e^{-k_{A \to B} t}$, where $k_{A \to B}$ is the rate coefficient. If each $w_{A_n \to B}(t)$ is (1) independent of n and (2) has the exponential form then their average is also obviously exponential and so the first-order kinetics, indeed, applies to the collective states. This can happen if the network in question possesses very special properties where, for example, the rate coefficient for making a transition from any of the A substates to any of the B substates is the same. It is unlikely that this would be a common scenario. However, another possibility exists, where the transitions among the substates of A are much faster than the overall transition from A to B. If this is the case then the memory of the initial state A_n will be lost long before a typical transition to B takes place. As a result, the likelihood of making a transition to B (per unit time, conditional upon being in A) will be independent of the internal state of A or the system's past, in accord with the assumptions that have led us to a derivation of first-order kinetics in Chapter 3.

To illustrate these arguments more quantitatively, consider the simple kinetic scheme shown in Fig. 5.6, right. It involves two A-states and one B-state. When in A, the molecule can jump between the two substates, a process characterized by a rate coefficient k_0. It can make a transition to B from either A_1 or A_2, with the corresponding rate coefficients equal to k_1 and k_2. Suppose at $t = 0$ the molecule is found in a particular substate of A, say A_1. To analyze the kinetics of transitions to B, it is helpful to consider the survival probability $w_s(t)$, i.e., the probability that the system is still in either of the two A-states at time t. Since $-dw_s = w_s(t) - w_s(t + dt)$ is the probability of making a transition to B between t and $t + dt$, the survival probability is related to $w_{A_1 \to B}(t)$ by

$$w_{A_1 \to B}(t) = -dw_s/dt,$$

and, in the case of first-order kinetics, would have to be equal to $\exp(-k_{A \to B} t)$. For our 3-state model, the survival probability can be estimated by solving the differential equations describing the probabilities $w_1(t)$ and $w_2(t)$ to populate the A substates,[5] which are given by:

$$dw_1/dt = -k_0 w_1 + k_0 w_2 - k_1 w_1$$
$$dw_2/dt = -k_0 w_2 + k_0 w_1 - k_2 w_2.$$

Solving these equations, with the initial conditions $w_1(0) = 1$, $w_2(0) = 0$, is somewhat tedious but straightforward. The survival probability is then found as the sum of the probabilities to be in either A-state,

$$w_s(t) = w_1(t) + w_2(t).$$

The reader may verify that the result can be written in the form

$$w_s(t) = \frac{k_1 - k_2 + \tilde{k} - 2k_0}{2\tilde{k}} e^{-(2k_0 + k_1 + k_2 + \tilde{k})t/2} + \frac{k_2 - k_1 + \tilde{k} + 2k_0}{2\tilde{k}} e^{-(2k_0 + k_1 + k_2 - \tilde{k})t/2},$$

(5.31)

where

$$\tilde{k} = \sqrt{4k_0^2 + (k_1 - k_2)^2}.$$

(5.32)

Since Eq. 5.31 contains two exponentially decaying terms rather than one, it does not describe a first-order transition from A to B. However the first of the two terms becomes vanishingly small if

$$|k_1 - k_2| \ll k_0.$$

(5.33)

---

[5] Note that, although the 3-state scheme of Fig.5.6 ignores the possibility of transitions from B to A, the survival probability of A would be unaffected by such transitions. Indeed, as shown in Chapter 3, the survival probability of a state is not affected by transitions into this state.

Indeed, applying Taylor expansion to Eq. 5.32, we find

$$\tilde{k} \approx 2k_0 \left[1 + \frac{(k_1 - k_2)^2}{8k_0^2} \right] \approx 2k_0.$$

Substituting this into Eq. 5.31 we find that the first term is negligible. The survival probability can now be estimated as the second term, which can be further approximated by

$$w_s(t) \approx e^{-\frac{k_1 + k_2}{2} t}.$$

Same result is obtained if one starts from A_2 so the survival probability is independent of the initial state. We are therefore dealing with an effectively first-order transition from A to B, with $k_{A \to B}$ given by the average rate coefficient,

$$k_{A \to B} \approx \frac{k_1 + k_2}{2}.$$

Let us now note that the assumption expressed by Eq. 5.33 encompasses two distinct physical scenarios. In the first, both k_1 and k_2 are much smaller than k_0. In this case, the interconversion within the collective state A takes place much faster (with a characteristic timescale of k_0^{-1}) than the transitions to B (a characteristic timescale of $\frac{2}{k_1 + k_2}$). In accord with our arguments above, this separation of timescales ensures that transitions from A to B happen according to first-order kinetics, with a rate coefficient that is averaged over the two possible transition pathways. In the second scenario, the individual rate coefficients do not have to be small as compared to k_0; however their difference is small. In other words, the rate coefficients for the two transition pathways are approximately equal, $k_1 \approx k_2$. If so, the internal state within A has no effect on the transition rate and first-order kinetics is recovered again.

Although the origins of first-order kinetics in complex systems can often be traced to the existence of slow transitions associated with high activation barriers, other scenarios exist where no individual transition is slow, yet first-order kinetic laws apply. To illustrate some of rather subtle mathematical issues encountered when analyzing complex kinetic networks, consider the scheme shown in Fig.5.7, the top. It describes a one-dimensional random walk, with N states on the left designated as A-states and the ones on the right as B-states. Can the dynamics of transitions from the collective state A to B be approximated as first-order kinetics? To answer this question, we need to find out how sensitive the time to reach B is to the starting point in A. Let $t_{n \to B}$ be the mean time it takes to reach B having started in A_n and $t_{n \to n'}$ the mean time to reach $A_{n'}$ having started in A_n, where $n' > n$.[6] Since any random walk leading from A_n to A_N will necessarily visit $A_{n'}$, we can claim that $t_{n \to B}$ is the sum of the time to reach $A_{n'}$ and the time to go from $A_{n'}$ to B,

$$t_{n \to B} = t_{n \to n'} + t_{n' \to B}. \tag{5.34}$$

This equation necessitates that $t_{n \to B}$ increases as n is decreased. This conclusion is, of course, not surprising: It just states that the further the system starts away from B

---

[6] Such times are examples of mean first passage times discussed in the next chapter.

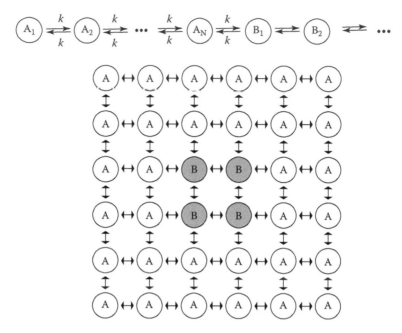

FIGURE 5.7 Kinetic networks where getting to the collective state B involves one-dimensional (top) and two-dimensional (bottom) random walks.

the longer it takes to get there. Importantly, however, the n dependence of the time turns out to be quite strong. It can be shown (see the next section) that

$$t_{N \to B} = \frac{N}{k}, \tag{5.35}$$

and, analogously,

$$t_{N-1 \to N} = \frac{N-1}{k}.$$

Then, using the above additivity of times, we find

$$t_{N-1 \to B} = \frac{N}{k} + \frac{N-1}{k} = \frac{2N-1}{k},$$

$$t_{N-2 \to B} = \frac{3N-3}{k}$$

$$\cdots$$

$$t_{N-n \to B} = \frac{(n+1)(N-n/2)}{Nk},$$

$$\cdots$$

$$t_{1 \to B} = \frac{N(N+1)}{2k}. \tag{5.36}$$

Clearly, the n-dependence is non-negligible under any conditions. In particular, the times $t_{N \to B}$ and $t_{1 \to B}$ differ by a factor of $(N+1)/2$. This indicates that a first-order kinetics model for the transitions between A and B would be inapplicable in this case and that hopping between the collective states A and B is a highly non-Markovian process with strong memory of the starting state.

The calculation above implies that the system does not have enough time to "forget" its initial state in A before it makes a transition to B. Curiously, a seemingly small modification of the model may alter the situation significantly. Imagine, for example, that the random walk of Fig.5.7 is biased and all rate coefficients for going left are greater than those for going right. A sufficiently strong bias would then drive the system left toward the most probable state A_1, which would be likely reached much sooner than there is a chance of a transition to B. Any transition to B would then require a long and rare excursion in the unfavorable direction. As a result, the time to reach B would be almost independent of the initial state within A and would further be close to $t_{1 \to B}$. Again, the reader has likely recognized the "time-scale separation" requirement here: If the time to reach the product B is much longer than the time to forget the initial condition within A then the transition from A to B can be approximated by first-order kinetics.

Exercise

By solving the following problem, you will quantify the above handwaving arguments. Let us modify the one-dimensional random-walk scheme of Fig.5.7 by replacing each rate coefficient k by k_+ if the step occurs from left to right and by k_- otherwise. Derive the analogs of Eqs.5.35-5.36 in this case. To do so you may need to adapt the derivation of Eqs.5.35 from the next section. Alternatively, you may take the following result for granted:

$$t_{N \to B} = k_+^{-1}(1 + \xi + \xi^2 \ldots + \xi^{N-1}),$$

where $\xi = k_-/k_+$. The time $t_{n \to B}$ to reach B from a starting point A_n other than A_N can now be calculated using the additivity rule (Eq. 5.34), as above. Now investigate the dependence of the time $t_{n \to B}$ on the location of the starting point n, the length of the random walk N, and the ratio ξ. Under what conditions does this time become weakly dependent on n?

An even more interesting (and less intuitive) situation arises if we consider kinetic networks in higher-dimensional spaces. Imagine, for example, that the states of a molecule can be arranged on a two-dimensional square lattice (as shown in Fig. 5.7, the bottom) or, analogously, on a three-dimensional cubic lattice (not shown). A subset of states are designated to be the B-states while the rest of the states belong to A. All transitions between adjacent lattice sites have the same value of the rate coefficient k. It turns out that whether or not the overall kinetics of transitions from A to B can be described by a first-order kinetic scheme fundamentally depends on the dimensionality of the lattice space. To illustrate this dependence, imagine that all of the A-states are confined within a d-dimensional cube (which is a line segment in 1D and a square in 2D) with an edge length of N (where the unit of length corresponds to a single transition between neighboring states). Let us further assume that the number of B-states is small as compared to the cube volume N^d. To reach B from a state in A a typical distance of $\sim N$ must be traveled. Since the motion we consider is a random walk, this will actually take a number of steps n that is much higher than N.

Specifically, using the results of Appendix A, we expect

$$n \sim N^2. \tag{5.37}$$

At the same time, the total number of states lying within the volume "explored" by the random walker would be

$$n_{states} \sim N^d.$$

If N is large and $d < 2$ then we have $n_{states} \ll n$. That is most of the states within the explored volume will be visited more than once. If, in contrast, $d > 2$ then the number of states within the cube far exceeds the number of steps n_{steps} made and, therefore, it also exceeds the number of states actually visited by the molecule. In the former case, the state B will be likely reached in just $\sim n$ steps. For example, in the case $d = 1$ (Fig. 5.7, top) the typical number of steps to reach B is $\sim N^2$. In fact, *every intervening state* between A_1 and B will be visited with certainty by the time the system arrives in B. Therefore, the time to reach any target B in 1D is of order N^2, where N is the number of A-states. On the other hand, Eq. 5.37 can also be interpreted as the number of steps it takes the system to forget its initial condition, as the system can now be found anywhere within the cube.[7] Therefore the time to reach B is comparable to the time to forget where the system has started. As the two times are of similar magnitude, no timescale separation exists and the A-to-B transition cannot be described by first order kinetics.

In contrast, when $d > 2$, it is unlikely that the state B will be reached in just n steps (given by Eq. 5.37), as this number is much smaller than the number of states within the cube. For example, by the time a 3D random walker traverses the distance from one vertex of the cube to another (which takes $\sim n$ steps), only a small fraction of cube states will be visited. Thus the number of steps (or the corresponding time) required to finally reach B is much greater than the number of steps n (or the corresponding time) required to forget the initial state. When such time separation exists, we expect the A-to-B transitions to be well described by first-order kinetics. We note that a random walk on a plane ($d = 2$) is a borderline case between the two limits. This case should be treated with special caution but will be simply avoided here.

Why should anyone be interested in random walks in multidimensional spaces? One good reason is that, for $d = 3$, we have effectively developed a discrete model of 3D diffusional search for a small target and showed that a simple kinetic model treating A and B as collective states is justifiable in this case. Such search processes, also referred to as diffusion controlled reactions, are important in chemical physics and, especially, in biophysics. For example, biological function of many proteins (e.g., enzymes) commonly requires that they find their cellular targets via diffusive search. Interestingly, this can happen through a 3D search (the protein diffuses in solution), 1D search (e.g., the protein slides along the DNA chain), or a combination of the two. Dimensionality considerations are clearly important in such situations.

The description of a protein that undergoes internal dynamics may further require more degrees of freedom than the x, y, and z coordinates describing its location in space. As a result, the effective dimensionality of space explored by proteins or

---

[7] Multiplied by the average duration of a step, this quantity is the time τ_{diff} from Section 4.6.

other molecules (particularly chain molecules, which can adopt numerous different conformations) often differs from 3. The crucial role of dimensionality in diffusive search problems involving chain molecules was first noted by de Gennes [4]. A highly entertaining and insightful introduction to the subject can be found in ref. [5].

5.9   FURTHER DISCUSSION: DERIVATION OF EQ. 5.35

Despite its simplicity, proving Eq. 5.35 is surprisingly tricky because a transition from A_N to its adjacent state in B may involve arbitrarily long excursions to other A-states. The simplest proof I am aware of employs the elegant trick described in ref. [6]. Suppose we start in the state A_N and follow the time evolution of the system until it makes a transition to B. The survival probability $w_s(t)$ that no transition to B has occurred is given by

$$w_s(t) = w_1(t) + w_2(t) + \cdots + w_N(t), \tag{5.38}$$

where the individual probabilities to be in each state obey the equations

$$dw_i/dt = kw_{i-1} + kw_{i+1} - 2kw_i,$$

for $i = 2, \ldots, N - 1$. For $i = 1$ we have

$$dw_1/dt = kw_2 - kw_1,$$

while the evolution of w_N involves an irreversible step describing the transition to B,

$$dw_N/dt = kw_{N-1} - 2kw_N.$$

Now integrate each of these equations over time from 0 to ∞. The left hand side of each equation gives

$$\int_0^\infty dt(dw_i/dt) = w_i(\infty) - w_i(0).$$

Our initial conditions imply that $w_N(0) = 1$ and $w_i(0) = 0$ for $i \neq N$, while $w_i(\infty) = 0$ because our system will reach the state B (and thus leave the state A) sooner or later. Introducing

$$\tau_i = \int_0^\infty dt\, w_i(t) \tag{5.39}$$

we find

$$0 = k\tau_2 - k\tau_1, \tag{5.40}$$
$$0 = k\tau_{i-1} + k\tau_{i+1} - 2k\tau_i \tag{5.41}$$

for $1 < i < N$ and, finally,

$$-1 = k\tau_{N-1} - 2k\tau_N. \tag{5.42}$$

If we add up Eqs.5.40 through 5.42 we will find that $k\tau_N = 1$ or $\tau_N = 1/k$. Now these equations can be easily solved to give:

$$\tau_1 = \tau_2 = \ldots \tau_N = 1/k. \tag{5.43}$$

Let $w_{N \to B}(t)$ be the probability density for the time t to arrive in B (having started in A_N). Since transitions to B are the only possible cause of a change in the survival probability w_s, we have

$$dw_s = w_s(t+dt) - w_s(t) = -w_{N \to B}(t)dt$$

or

$$w_{N \to B}(t) = -dw_s/dt.$$

The mean time to arrive in B can now be calculated as

$$t_{N \to B} = \int_0^\infty dt\, t\, w_{N \to B}(t) = \int_0^\infty dt\, w_s(t),$$

where integration by parts was used. Finally, recalling Eqs.5.38, 5.39, and 5.43, we obtain

$$t_{N \to B} = N/k.$$

Exercise

Find the expression for $t_{n \to B}$ using the same method but starting from an arbitrary state A_n belonging to A.

REFERENCES

1. Robert Zwanzig, *Nonequilibrium Statistical Mechanics*, Oxford University Press, 2001.
2. Peter Hänggi, Peter Talkner, and Michal Borkovec, "Reaction-rate theory: fifty years after Kramers," *Reviews of Modern Physics*, vol. 62, pp. 252-342, 1990.
3. Daan Frenkel and Berend Smit, *Understanding Molecular Simulation*, Academic Press, 2002.
4. P.G. de Gennes, "Kinetics of diffusion controlled processes in dense polymer systems. I. Nonentangled regimes",*J. Chem. Phys.*, vol. 76, p. 3316, 1982.
5. A. Grosberg, "A few disconnected notes related to Levinthal paradox", *Journal of Biomolecular Structure and Dynamics*, vol. 20, p. 317, 2002.
6. Jianshu Cao, "Michaelis-Menten equation and detailed balance in enzymatic networks", *J. Phys. Chem. B*, vol. 115, p. 5493, 2011.

6 Molecular Transition Paths: Why Going Uphill May Be Faster

If the dynamical laws of an isolated molecular system are reversible the kinetic theory requires that in the long run every type of motion must occur just as often as its reverse, because the congruence of the two types of motion makes them *a priori* equivalent. This implies that if we wait a long time so as to make sure of thermodynamic equilibrium, in the end every type of motion is just as likely to occur as its reverse. One consequence of this *principle of dynamical reversibility* is the condition that when a molecule changes a certain number of times per second from the configuration A to the configuration B the *direct* reverse transition B ⟶ A must take place equally often....

Lars Onsager, *Reciprocal relationships in irreversible processes*

A molecule involved in the reaction

$$A \underset{k_{B \to A}}{\overset{k_{A \to B}}{\rightleftharpoons}} B$$

spends most of its time as either A or B. As discussed in the previous chapter (cf. Fig. 5.2), the actual transition events are often very short, so short that, for most practical purposes, we can replace the more detailed molecular trajectory shown in Figure 5.2 by a simplified, two-state process shown in Figure 3.4. No matter how short, however, the transition events are definitive: It is during those rare excursions that the molecule visits the regions of its potential energy surface that—ultimately—determine the values of the rate coefficients $k_{A \to B}$ and $k_{B \to A}$.

To observe what happens during those transitions, single-molecule measurements (as opposed to ensemble measurements) are imperative. Indeed, as discussed in Chapter 3, an ensemble observation of any time-dependent molecular property requires the ensemble members to be synchronized. Although it is possible to prepare all the molecules in the same initial state, it is impossible to make them all jump to another state at the same time as we have no control over the random waiting time before the jump happens.[1]

In practice, single-molecule observation of the details of a single transition presents a great experimental challenge because of its very short duration. Optical

---

[1] Using clever experimental methods employing short laser pulses, it is sometimes possible to prepare an ensemble of molecules that are all initially halfway between A and B. Because this halfway state is unstable, the molecule will almost immediately proceed towards either A or B, thus enabling an ensemble observation of "half-transitions" to either state. A disadvantage of this approach is that the initial ensemble created this way is generally different from the ensemble of molecules crossing the transition state under equilibrium conditions and so it may not be representative of transitions happening in the absence of the laser.

measurements, for example, will not resolve any temporal details that take place over timescales shorter than the time interval between photons emitted by the molecule. Recently, William A. Eaton and his colleagues have succeeded in measuring, through a clever analysis of individual photons,[2] the duration of individual transition events in the folding and unfolding of proteins [1]. Measurement of the time the molecule spends in transit between two stable states is an important initial step towards mechanistic understanding of molecular transition pathways. In what follows, I will discuss some of the peculiar properties of this time.

Exercise

Suppose the transition from A to B involves a single intermediate state I. In other words, the molecule's kinetics can be described by the following scheme:

$$A \underset{k_{I \to A}}{\overset{k_{A \to I}}{\rightleftharpoons}} I \underset{k_{B \to I}}{\overset{k_{I \to B}}{\rightleftharpoons}} B.$$

What is the mean duration of the transition $A \longrightarrow I \longrightarrow B$ in which the system starts in A and proceeds directly to B without ever returning to A?

6.1 TRANSIT TIMES VS. FIRST PASSAGE TIMES

First of all, we need to define what we mean by the duration of a transition event, or, what we will call it, transit time. The kinetics of a chemical reaction is measured by monitoring some quantity s (which we call the reaction coordinate) as a function of time. As before, it is helpful to have in mind a specific model describing $s(t)$. This can, for example, be the motion in some effective potential $V(s)$ (Figure 6.1), governed by a Langevin equation

$$m\ddot{s} = -V'(s) - \eta\dot{s} + R(t). \tag{6.1}$$

We define a transition region by specifying its two boundaries, s_A and s_B. A successful transition from A to B begins when the molecule crosses the boundary s_A going towards B. It ends when the molecule exits the transition region $s_A < s < s_B$ by crossing the boundary s_B. The time t_{AB} it takes to travel across the transition region is what we call the transit time and the path $s(t)$ followed during this time is a transition path. For example, the red trajectory segments in Figure 6.1 are transition paths from A to B and from B to A. When measuring transit times, only successful transitions count. That is, any event where the molecule enters the transition region through one boundary and exits through the same boundary, having failed to reach the other state, does not contribute to a transit time.

Transit times generally depend on the choice of the boundaries. We will, however, see that this dependence is fairly weak. Because most molecular configurations

---

[2] This single-photon statistical analysis was developed by Irina Gopich and Attila Szabo [2] and will be further discussed in Chapter 7.

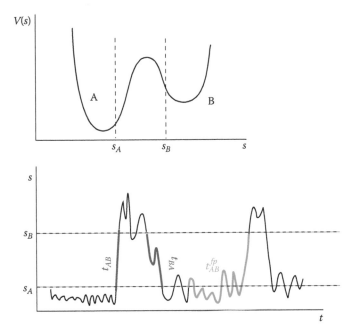

FIGURE 6.1 The transition region separating two stable states, A and B, is defined by two boundaries, s_A and s_B. A transition path (red) enters the transition region through one boundary and exits it through the other. The duration of the transition path is the transit time. In contrast, the first passage time is the time it takes to arrive at one boundary, having started from the other. The system may exit and reenter the transition region during such an event, as shown in green.

corresponding to A or B are concentrated within a few $k_B T$ from the bottom of the respective potential well, a good physical choice of the boundaries is such that the energy $V(s_A)$ or $V(s_B)$ is just a few $k_B T$ above the bottom of the respective well. This ensures that, in the scenario shown in Figure 6.1, most A configurations will be found to the left of s_A and most B configurations will lie to the right of s_B, while the transition region itself will be sparsely populated.

It is also instructive to consider a different timescale. Starting with a crossing of the boundary s_A we follow the trajectory $s(t)$ until it arrives, for the first time, to s_B. This event is illustrated in Figure 6.1 in green. We call the duration of this trajectory segment the first passage time, t_{AB}^{fp}, from A to B. Any transition path, of course, also results in a first passage event. However, the converse is not true: A first passage time to B may include numerous failed attempts to make a transition to B, where the system exits and reenters the transition region through s_A. As a result, we expect the mean first passage time to be longer than the mean transit time.

A good estimate of the mean first passage time can be obtained using the two-state model of reaction kinetics discussed in Chapter 3. If the molecule starts, at some moment, in A, the mean time it takes to cross over to B is simply $1/k_{A \to B}$. Although the two-state model does not account for the mean first passage dependence on the precise boundaries, it seems sensible to propose that this dependence is weak so that

$$\langle t_{AB}^{fp} \rangle \approx 1/k_{A \to B}.$$

On the other hand, the two-state model cannot give a meaningful estimate of the transit time, as it predicts the duration of a single jump between two states to be infinitely short.

Exercise

(A) Is it true that the mean transit times from s_A to s_B, from s_B to s_C, and from s_A to s_C, where $s_A < s_B < s_C$, satisfy the equation

$$\langle t_{AC} \rangle = \langle t_{AB} \rangle + \langle t_{BC} \rangle?$$

(B) Is the analogous equation valid for the mean first passage times,

$$\langle t_{AC}^{fp} \rangle = \langle t_{AB}^{fp} \rangle + \langle t_{BC}^{fp} \rangle?$$

6.2   TIME REVERSAL SYMMETRY AND ITS CONSEQUENCES FOR TRANSIT TIMES

Here I will prove that the mean transit time from A to B is the same as from B to A, i.e.,

$$\langle t_{AB} \rangle = \langle t_{BA} \rangle.$$

This may seem counterintuitive. In Figure 6.1, for example, the transition from A to B is uphill in energy and from B to A is downhill. It would seem that going uphill should be slower! This expectation, however, turns out to be wrong. It is true that the mean first passage time in the uphill direction, $\langle t_{AB}^{fp} \rangle$, would likely be longer[3] than $\langle t_{BA}^{fp} \rangle$. But the mean duration of a successful barrier crossing event is the same in each case.

To prove this fact, we first need to discuss an important property of a molecular trajectory: its time reversal symmetry. If $s(t)$ satisfies Newton's second law

$$m\ddot{s} = -V'(s), \tag{6.2}$$

then so does its time-reverse

$$\tilde{s}(t) = s(-t), \tag{6.3}$$

since, obviously, $\ddot{\tilde{s}} = \ddot{s}$. Langevin trajectories are, however, stochastic[4] and so their time reversal symmetry must be restated in terms of their statistical properties. Consider a trajectory of a Langevin particle such as one shown in Fig. 4.6. If someone

---

[3] To be precise, this is true if the reaction is uphill in free energy rather than just energy.
[4] At zero temperature, the noise $R(t)$ disappears and the Langevin equation becomes deterministic. It also loses its time reversal symmetry as it predicts the molecule to slow down and eventually to stop moving. The zero-temperature case is pathological and we need not worry about it: After all, laws of quantum mechanics, rather than classical trajectories, have to be used in this case.

gave us two plots, one of $s(t)$ and the other of $s(-t)$ without telling us which is which, it would be impossible for us to establish which one runs forward in time. The rigorous proof of this statement is outside the scope of this book. However I am going to try to convince you that this is plausible using the following argument. Recall that our reason for modifying Newton's equations of motion was to account for the effect of other molecules that we did not want to consider explicitly [4]. The underlying dynamics of the extended system that includes other molecules is still Newtonian and, therefore, time reversible.[5]

Going back to transit times, consider a long trajectory $s(t)$ (cf. Fig.6.1). If it is time-reversed, every transition path from A to B will become a transition path from B to A and vice versa. But we know that $s(t)$ and $s(-t)$ have indistinguishable statistical properties. It follows immediately that t_{AB} has exactly the same probability distribution as t_{BA} and so their mean values are also the same.

This argument does not apply to first passage times. A time-reversed first passage event from A to B (e.g., the green trajectory segment in Fig.6.1) is not necessarily a first passage event from B to A. Therefore there is no reason for $\langle t_{BA}^{fp} \rangle$ to be equal to $\langle t_{AB}^{fp} \rangle$.

Let me conclude this section with a comment. If $k_{B \to A} \gg k_{A \to B}$, it is often said that a transition from A to B is much less likely to occur than from B to A. One, however, has to be careful. If we watch the system's evolution over a long period of time, we will count equal number of transitions from A to B as from B to A, since each A-to-B transition is followed by a B-to-A one. In this regard both types of transitions occur with identical frequencies. It is true, though, that much more time will be spent in state A than in state B and so the probability, per unit time, to make a transition from A to B (provided one is already in A) is lower than the probability of the reverse event. It is thus important to keep in mind that the rate coefficients are conditional probabilities.

6.3   TRANSIT TIME THROUGH A PARABOLIC BARRIER

In Newtonian dynamics, the initial position and velocity at $t = 0$ completely determine the fate of the system at $t > 0$. Since Langevin dynamics is stochastic, those initial conditions do not completely specify the future trajectory but, nevertheless, the past history of the system at $t < 0$ affects its future only through the values of $s(0)$ and $\dot{s}(0)$. It then follows that the manner in which our molecule evolved outside the transition region before arriving at its boundary s_A does not affect any properties of the transit time t_{AB}. Therefore, t_{AB} depends on the properties of the potential $V(s)$ inside (but not outside) the transition region. Again, this is different from the case of

---

[5] This argument appears to fail when applied to the macroscopic world, where irreversible phenomena are abundant. We are, for example, quite sure that a car will slow down rather than speed up every time we apply the brakes. This apparent paradox is resolved if the timescales necessary to observe the reversal of apparently irreversible events are considered, which turn out to be astronomic for systems that consist of many molecules. The origins of irreversibility and its reconciliation with time-reversal symmetry of underlying physical laws is discussed, for example, in Richard Feynman's famous lectures on the character of physical law [3], which are strongly recommended to the interested reader.

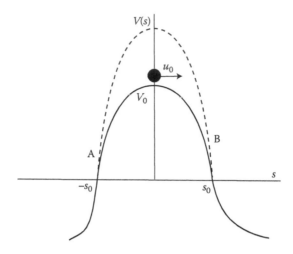

FIGURE 6.2 Transitions over a symmetric parabolic barrier. In the case where energy dissipation during a transition event is negligible, typical transit times across a higher barrier are shorter than those for a lower barrier. The transit times do not depend on the shape of the potential outside the transition region $-s_0 < s < s_0$.

first passage times, which depend on the shape of $V(s)$ both inside and outside the transition region.

Here we will estimate the transit time for the simple case where the potential energy inside the transition region can be approximated by a quadratic function,

$$V(s) = V_0 - \frac{m}{2}\omega_b^2 s^2,$$

where ω_b is the upside-down barrier frequency. The transition region is defined as $s_B = -s_A = s_0$. The potential is shown in Fig. 6.2. We will further assume that the energy dissipated through friction during a single transition event is negligible and so the shape of the transition paths can be adequately described by Newton's second law, Eq. 6.2, which gives:

$$\ddot{s} = \omega_b^2 s.$$

Any solution to this equation is of the general form

$$s(t) = ae^{\omega_b t} + be^{-\omega_b t}.$$

Without loss of generality, we can assume that the trajectory crosses the top of the barrier at $t = 0$, i.e., $s(0) = 0$ and so $a = -b$ and

$$s(t) = a[e^{\omega_b t} - e^{-\omega_b t}]. \tag{6.4}$$

In order to cross the barrier, the molecule's energy must be higher than the barrier height. Let u_0 be the velocity right at the moment when the barrier is being crossed. The total energy is then $E = V_0 + mu_0^2/2$. A trajectory with any positive value of u_0 is a transition path. High values of u_0 are, however, improbable, which is reflected in a very small value of the corresponding Boltzmann exponential $\exp(-E/k_B T)$. As this exponential falls off with a characteristic energy scale $k_B T$, most likely transition paths are those whose energy is within $\sim k_B T$ from the top of the barrier, i.e.,

$$mu_0^2 \sim k_B T$$

and

$$u_0 \sim \sqrt{k_B T / m}. \tag{6.5}$$

Using Eq. 6.4 and demanding that $\dot{s}(0) = u_0$, we obtain

$$a = \frac{u_0}{2\omega_b} \sim \frac{\sqrt{k_B T / m}}{\omega_b}.$$

Since any numerical factors in these crude estimates are questionable, those are dropped here and throughout the rest of this section.[6] To transit time t_{AB} is now the time it takes the molecule to climb to the top of the barrier starting from $s_A = -s_0$ plus the time it takes to slide from the top of the barrier down to $s_B = s_0$. It is easy to see that these two times are equal so t_{AB} is twice the time to go from $s(0) = 0$ to $s = s_0$, having started with an initial velocity u_0. To find this time, we need to solve the equation

$$s(t) = s_0,$$

where $s(t)$ is given by Eq. 6.4. If the solution t of this equation also satisfies the inequality $\omega_b t \gg 1$, then the exponential $\exp(-\omega_b t)$ can be further neglected and we get

$$ae^{\omega_b t} \sim s_0$$

and

$$t = \omega_b^{-1} \ln \left(\frac{s_0 \omega_b}{\sqrt{k_B T / m}} \right).$$

Finally,

$$t_{AB} \sim 2t = \omega_b^{-1} \ln \left(\frac{m \omega_b s_0^2}{k_B T} \right) \sim \omega_b^{-1} \ln \left(\frac{V_0}{k_B T} \right), \tag{6.6}$$

where $V_0 = m \omega_b s_0^2 / 2$ is the barrier height relative to the entrance to the transition region. The above assumption $\omega_b t \gg 1$ is justified if the barrier is high enough, $V_0 \gg k_B T$. Notice that if ω_b is constant, the resulting barrier height dependence of the transit time is rather weak (logarithmic), in contrast to the strong, exponential barrier height dependence of the transition rate coefficient. Indeed, two proteins, whose folding rates differed by many orders of magnitude, were found to display very similar transit times [1].

Let us now compare two barriers of the same width, as in Fig.6.2. The barrier height is then controlled by the upside-down frequency ω_b, which would be higher for the higher barrier. It is then easy to see from Eq. 6.6 that the transit time for the higher barrier will be shorter. This result appears counterintuitive: Why would climbing a higher barrier be faster? The explanation is simple: To overcome a taller

---

[6] A more accurate solution that does not make any of the above approximations can be found in ref. [5].

barrier, a higher initial kinetic energy is required and so the successful transition paths happen to cross the barrier region faster. In this regard, again, transit times behave very differently from mean first passage times, as the latter grow if the barrier is increased.

Exercise

Assuming Newtonian dynamics in a one-dimensional potential, prove that the flat potential (i.e., constant $V(s)$) maximizes the mean transit time between any two points s_A and s_B.

6.4 FURTHER DISCUSSION: HOW TO FOLLOW A LANGEVIN TRAJECTORY BACKWARD IN TIME

When discussing the computation of rate coefficients in the preceding chapter, a useful trick was employed. If we want to study how a molecule crosses an energy barrier, we can place it at the top of the barrier (say at $s = 0$) and follow its dynamics both forward and backward in time. By gluing the two trajectory pieces together we generate the entire transition path. This trick avoids the problem of waiting for a long time (i.e., a first passage time) before a successful transition occurs. To use this trick in conjunction with Langevin dynamics, we need to know how to run Langevin trajectories backward in time.

Retracing the past of a Newtonian trajectory generated by Eq. 6.2 is simple. If the position s and the velocity $\dot{s}$ are known at $t = 0$, the prior values of $s(t)$ can be recovered by simply reversing the velocity at $t = 0$, i.e., $\dot{s}(0) \rightarrow -\dot{s}(0)$. Indeed, $s(-t)$ obeys the same equation of motion as $s(t)$ but has the opposite value of the velocity $\dot{s}(0)$. If, for example, the velocity of a projectile is reversed, it will follow its own trajectory backward, provided that any frictional forces caused, e.g., by interaction with air, are negligible. But since the Langevin equation accounts for friction, the issue of retracing Langevin trajectories backward in time is more tricky.

Let us see what happens if we make the substitution $\tau = -t$ in Eq. 6.1. This results in the following equation:

$$m\ddot{s} = -V'(s) + \eta\dot{s} + R(-\tau). \tag{6.7}$$

Unlike the Newtonian case, the result does not look like the original equation. Rather, it appears that the sign of the friction coefficient has changed. The effectively negative friction spells trouble because it means that the "friction force" no longer acts in the direction opposite the motion but rather acts so as to increase the kinetic energy. In fact, if you treat $R(-\tau)$ as random and attempt to solve Eq. 6.7 numerically, the velocity and position will increase indefinitely until your computer outputs an error message. The problem is that $R(-\tau)$ can no longer be treated as random. To see this, suppose that the past trajectory $s(t)$ has been generated, numerically, by solving Eq. 6.1 for $-t_0 < t \leq 0$. In doing so, the computer's random number generator has produced some noise sequence $R(t)$. If we reverse the velocity at $t = 0$ and now solve the negative friction Langevin Equation 6.7 using exactly the same noise sequence but in reverse order, $R(-t)$, the past trajectory will be backtracked, just as in the frictionless case. But if, instead of using the precomputed sequence $R(-t)$, we use a newly generated sequence, the most likely outcome will be a trajectory that quickly goes out of control and gains an energy much higher than $k_B T$.[7] Further insight into this negative friction scenario can be gained if we think, again, of both the friction force and the noise $R(t)$ as a result of the system's interaction with the surrounding molecules. To make the system backtrack its own trajectory, it is necessary to reverse not only the velocity of the system itself but also the velocity of each surrounding molecule. If we could succeed in this unrealistic endeavor, the force acting on the system at each moment t would then be exactly the same as its past value at $-t$, precisely as predicted by Eq. 6.7.

The "negative Langevin equation" is not very useful since it only works if we already know the molecule's past. We can, however, take advantage of the time reversal symmetry of the Langevin trajectories $s(t)$[8] as follows: First, we solve the Langevin equation,

$$m\ddot{\tilde{s}} = -V'(\tilde{s}) - \eta\dot{\tilde{s}} + R(t),$$

which has a *positive* friction coefficient and a randomly generated noise $R(t)$, forward in time with the reversed-velocity initial condition

$$d\tilde{s}/dt|_{t=0} = -ds/dt|_{t=0}$$

that will ensure continuity of the velocity $\dot{s}(t)$ at $t = 0$. Second, we simply calculate $s(t < 0)$ by setting $s(t) = \tilde{s}(-t)$. Naturally, since Langevin dynamics is stochastic, we cannot claim that this recovers *the* past of the Langevin system at $t < 0$, but the time-reversal symmetry ensures that a *statistically valid* trajectory is produced.

REFERENCES

1. Hoi Sung Chung, Kevin McHale, John M. Louis, and William A. Eaton, "Single-molecule fluorescence experiments determine protein folding transition path times," *Science*, vol. 335, pp. 981-984, 2012.
2. Irina V. Gopich and Attila Szabo, "Decoding the pattern of photon colors in single-molecule FRET", *J. Phys. Chem. B*, vol. 113, 10965-10973, 2009.
3. Richard P. Feynman, *The Character of Physical Law*, Modern Library, 1994.
4. Robert Zwanzig, *Nonequilibrium Statistical Mechanics*, Oxford University Press, 2001.
5. Srabanti Chaudhury and Dmitrii E. Makarov, "A harmonic transition state approximation for the duration of reactive events in complex molecular rearrangements", *J. Chem. Phys.*, vol. 133, 034118, 2010.

---

[7] Such an amuck trajectory is still a valid member of the statistical ensemble but an improbable one, under thermal equilibrium conditions.

[8] At this point of discussion, we should appreciate that this symmetry is a highly nontrivial and even confusing result, given that, unlike in Newtonian dynamics, the transformation $t \rightarrow -t$ applied to a Langevin equation does not result in the same equation!

7 Properties of Light Emitted by a Single Molecule and What It Can Tell Us about Molecular Motion

Notwithstanding the complete experimental verification of the theory of diffraction, reflexion, refraction, dispersion, and so on, it is quite conceivable that a theory of light involving the use of continuous functions in space will lead to contradictions with experience, if it is applied to the phenomena of the creation and conversion of light.

A. Einstein, *On a heuristic point of view about the creation and conversion of light*

This chapter is concerned with the measurements where molecular properties are deduced from the light emitted by the molecule of interest. An individual molecule emits one photon at a time. The resulting sequence of photons (see Fig.7.1) can be recorded using a highly sensitive single-photon detector. The emission of photons is governed by the laws of quantum mechanics, according to which it is fundamentally impossible to know their exact arrival times. Rather, only the probability of emitting a photon during a given time period can be predicted. The statistical properties of the resulting random photon sequence are modulated by the motion of the emitting molecule. The challenge is then to "decode" this sequence and to uncover the underlying molecular motion. The following sections describe how this can be achieved. We will begin with a discussion of photon sequences emitted by non-single-molecule light sources, then move on to "static" single-molecule emitters whose properties do not change in time. Next, we will give an example of an experiment where photon statistics is affected by molecular dynamics and, finally, outline various strategies through which such dynamics can be inferred from the data.

7.1 POISSON PROCESS AND NONSINGLE-MOLECULE LIGHT SOURCES

The statistical properties of photon sequences originating from single-molecule emitters differ from those produced by generic light sources. Let us start with the latter. A typical light source (such as a lightbulb) is composed of numerous independent light-emitting atoms or molecules. A sufficiently sensitive photodetector may be used to record the arrival time of each individual photon. The simplest assumption one could adopt to describe this kind of measurement is that the arrival times for different photons are completely uncorrelated (see, however, the discussion at the end of this section). If the properties of the light source (such as, e.g., its brightness) do not

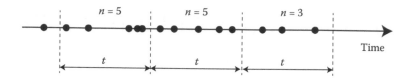

FIGURE 7.1 To analyze the statistics of photons arriving at a photodetector (shown as dark circles), we can break the time into bins of equal length t and measure the probability $w_n(t)$ to find n photons within a bin.

depend on time then we should expect that the probability of detecting a photon during an infinitesimal time interval dt is proportional to dt, with the proportionality constant being independent of time or the arrival times of other photons. Let us call this probability λdt, where λ is the probability of detecting a photon per unit time (which will be referred to as the photon count rate). To estimate λ we could measure the number of photons n detected over a certain period of time t and divide it by t:

$$\lambda \approx n/t. \tag{7.1}$$

But, since photon arrival times are random, the exact number of photons found during any finite time interval t will fluctuate around its mean value λt, as illustrated in Fig.7.1. A complete description of this process thus requires the probability $w_n(t)$ that a specified number of photons, n, is observed during the time interval t. To obtain this probability, it is helpful to think of the number of detected photons, n, as the "state" of the system at time t. At $t = 0$ we have $n = 0$. Each time a new photon is detected, the system undergoes a first-order irreversible "reaction" $n \to n + 1$. The overall process thus can be described by the kinetic scheme:

$$0 \xrightarrow{\lambda} 1 \xrightarrow{\lambda} 2 \xrightarrow{\lambda} 3 \xrightarrow{\lambda} 4 \to \cdots.$$

Accordingly, $w_n(t)$ is described by a system of differential equations of the form

$$dw_n(t)/dt = \lambda w_{n-1}(t) - \lambda w_n(t), n > 0 \tag{7.2}$$

and

$$dw_0(t)/dt = -\lambda w_0(t), \tag{7.3}$$

with the initial conditions

$$w_n(0) = 0, n > 0$$

and

$$w_0(0) = 1.$$

The solution to Eq. 7.3 is

$$w_0(t) = e^{-\lambda t}. \tag{7.4}$$

Substituting this into the differential equation for w_1 (Eq. 7.2) we can find the time dependence of $w_1(t)$, then substitute the result into the equation for $w_2(t)$ and so on. Suppose we have found $w_{n-1}(t)$ and now wish to solve for $w_n(t)$. It is convenient to use the ansatz

$$w_n(t) = \rho_n(t)e^{-\lambda t}.$$

Substituting this into Eq. 7.2, we obtain

$$(d\rho_n(t)/dt)e^{-\lambda t} = \lambda w_{n-1}(t),$$

which gives

$$w_n(t) = \rho_n(t)e^{-\lambda t} = \lambda e^{-\lambda t} \int_0^t dt' e^{\lambda t'} w_{n-1}(t').$$

In particular, using Eq. 7.4, we obtain

$$w_1(t) = \lambda e^{-\lambda t} \int_0^t dt' = \lambda t e^{-\lambda t}.$$

Continuing this process, it is easy to see that

$$w_n(t) = \lambda e^{-\lambda t} \int_0^t dt'(t')^{n-1}/(n-1)! = \frac{(\lambda t)^n}{n!} e^{-\lambda t}. \tag{7.5}$$

This result (Eq. 7.5) is known as the Poisson distribution and a process described by this distribution will be referred to as a Poisson process.

It is easy to check that the probabilities corresponding to different values of n add up to one:

$$\sum_{n=0}^{\infty} w_n(t) = e^{-\lambda t} \left[\sum_{n=0}^{\infty} \frac{(\lambda t)^n}{n!} \right] = 1$$

since the expression inside the square brackets is the Taylor expansion of $e^{\lambda t}$. It is further straightforward to show that the average number of photons detected over a time interval t is λt, as previously anticipated:

$$\langle n(t) \rangle = \sum_{n=0}^{\infty} n w_n(t) = e^{-\lambda t} \sum_{n=0}^{\infty} \frac{(\lambda t)^n}{(n-1)!} = e^{-\lambda t}(\lambda t) \sum_{n=1}^{\infty} \frac{(\lambda t)^{n-1}}{(n-1)!} = \lambda t. \tag{7.6}$$

To quantify how much n fluctuates around its mean value $\langle n \rangle = \lambda t$ we calculate the average of

$$\Delta n^2 = (n - \langle n \rangle)^2.$$

First, we notice that this average can be written as

$$\langle \Delta n^2 \rangle = \langle n^2 \rangle + \langle n \rangle^2 - 2\langle n \rangle \langle n \rangle = \langle n^2 \rangle - \langle n \rangle^2.$$

Second, consider the auxiliary quantity:

$$\langle n^2 \rangle - \langle n \rangle = \langle n(n-1) \rangle,$$

which can be evaluated similarly to Eq. 7.6

$$\langle n(n-1) \rangle = \sum_{n=0}^{\infty} n(n-1)w_n(t) = e^{-\lambda t}(\lambda t)^2 \sum_{n=2}^{\infty} \frac{(\lambda t)^{n-2}}{(n-2)!} = (\lambda t)^2 = \langle n \rangle^2. \tag{7.7}$$

We therefore arrive at the following important result:

$$\langle \Delta n^2 \rangle = \langle n^2 \rangle - \langle n \rangle^2 = \langle n \rangle. \tag{7.8}$$

If, for example, we were to estimate the count rate λ using Eq. 7.1, this would result in a typical error of $\Delta n \approx \sqrt{n}$ or a relative error of $\Delta n/n \approx \sqrt{n}/n = 1/\sqrt{n}$.

Another important property of the Poisson process is the probability distribution of the lag time between successive photons. Since the probability that no photon is detected in a time interval t is simply $w_0(t)$ given by Eq. 7.4, the probability of detecting the next photon between t and $t + dt$, provided that the last photon was detected at $t = 0$, is

$$w_0(t) - w_0(t + dt) = -dw_0 = \lambda e^{-\lambda t} dt$$

and so the probability density of the lag time between two successive photons is given by:

$$w_{lag}(t) = \lambda e^{-\lambda t}. \tag{7.9}$$

Exercise

Show that the combined photon sequence produced by two independent Poisson sources obeys a Poisson process with a count rate $\lambda = \lambda_1 + \lambda_2$ equal to the sum of the count rates of each source.

How good is the Poisson process as a model for light emission from an ordinary source? While often being a reasonable approximation, it happens to miss an essential piece of physics: Rather than being completely statistically independent, photons from non-single-molecule sources generally tend to have "super-Poisson" statistics. That is, they tend to "bunch" or arrive in clumps. At first glance this may sound like nonsense: If each of the many atoms in the light source emits photons independently, how can detecting one photon enhance the probability of seeing the next one? Indeed, if photons were bullets shot randomly (and independently) from many guns, their Poisson statistics would be well justified. Photons are, however, not bullets, they are quantum particles obeying non-classical statistics.[1] Proper derivation of the bunching effect from quantum principles is rather subtle and will not be pursued here, since our main focus is on single molecules rather than conventional light sources. The interested reader can find such a derivation in many quantum optics textbooks and ref. [1] is recommended here as an especially accessible yet comprehensive introduction to the subject. The bunching effect, however, follows almost trivially from the classical picture of light as oscillating electromagnetic waves. Indeed, the probability of photon detection at any given moment must be proportional to the instantaneous light intensity, which is proportional to the square of the wave amplitude and is therefore a continuous function of time. Unless the incident light can be described as a perfectly monochromatic (i.e., single-frequency) electromagnetic wave (which could only be a

---

[1] More specifically, Bose-Einstein statistics, which is responsible for their tendency to congregate in the same quantum state.

good approximation when the light is emitted by a laser), the instantaneous intensity fluctuates. It does so, in particular, when the total wave amplitude is a sum of the waves contributed by many independent atoms. There must, then, be a decorrelation timescale over which the memory of light intensity is lost. In quantum optics, this timescale is known as the coherence time. The precise magnitude of the coherence time depends on the spectral properties of the light source and is unimportant here: The key point is that it's generally nonzero. Two photons separated by a lag time that is substantially longer than the coherence time are statistically independent; however the arrival times are clearly correlated otherwise. It should be noted that the above argument makes no detailed assumptions regarding the precise properties of the electromagnetic waves emitted by each atom but it does require a classical description of light (which turns out to be incorrect when the light is emitted by just one atom). To show that this description inevitably entails super-Poisson (and never sub-Poisson) statistics where successive photons attract rather than repel each other, let us modify the above model of the Poisson process by assuming that the instantaneous count rate (proportional to the light intensity) is time dependent,

$$\lambda = \lambda(t).$$

The solution of Eqs. 7.2 and 7.3 now becomes

$$w_n(t) = \frac{\Lambda^n(t)}{n!} e^{-\Lambda(t)}, \tag{7.10}$$

where

$$\Lambda(t) = \int_0^t \lambda(t')dt'.$$

Exercise

Prove Eq. 7.10.

Now it is easy to find the mean number of photons registered during the time period t: Simply replace λt by $\Lambda(t)$ in Eq. 7.6, which gives:

$$\langle n \rangle = \Lambda(t). \tag{7.11}$$

Likewise, a straightforward modification of Eq. 7.7 results in

$$\langle n(n-1) \rangle = \langle n^2 \rangle - \langle n \rangle = \Lambda^2(t). \tag{7.12}$$

Until now we treated the count rate $\lambda(t)$ and the related quantity $\Lambda(t)$ as deterministic functions of time. That is, Eqs. 7.11 and 7.12 will predict the outcome of an experiment where the number n of photons is measured repeatedly within a time window from 0 to t in which $\lambda(t)$ retraces *exactly the same* time dependence each time. However the experimenter has no precise control of $\lambda(t)$ and a typical measurement would rather involve multiple time windows of the same length t but with different time dependences of the instantaneous count rate within each window. Therefore $\Lambda(t)$

should be viewed as a fluctuating quantity. Provided that the fluctuations of the count rate $\lambda(t)$ are described by a stationary process whose statistical properties do not depend on the starting time of a window, averaging over multiple time windows is equivalent to averaging over possible values of $\Lambda(t)$ corresponding to the same window (but different histories of $\lambda(t)$). Performing this averaging, we find

$$\langle n \rangle = \langle \Lambda \rangle,$$
$$\langle n(n-1) \rangle = \langle n^2 \rangle - \langle n \rangle = \langle \Lambda^2 \rangle,$$

and, finally,

$$\langle \Delta n^2 \rangle = \langle n^2 \rangle - \langle n \rangle^2 = \langle n \rangle + \langle \Lambda^2 \rangle - \langle \Lambda \rangle^2 = \langle n \rangle + \langle (\Lambda - \langle \Lambda \rangle)^2 \rangle \geq \langle n \rangle. \quad (7.13)$$

Unless the count rate λ is constant, in which case $\langle \Lambda^2 \rangle = \langle \Lambda \rangle^2$ and so Eq.7.8 is recovered, we find $\langle \Delta n^2 \rangle$ to be greater than $\langle n \rangle$. That is, given the same mean count rate $\langle n \rangle$, the distribution of n is broader and, therefore, more photons can be cramped into the same time interval t as compared to the Poisson statistics case. This is super-Poisson statistics (or "photon bunching") where photon arrival times are correlated so as to effectively attract each other. The opposite case, where $\langle \Delta n^2 \rangle < \langle n \rangle$ and where photons effectively repel one another, would correspond to photon "antibunching" or sub-Poisson statistics. However it is clear from Eq. 7.13 that sub-Poisson statistics cannot be derived from the model of emission with a fluctuating count rate λ no matter how $\lambda(t)$ behaves. Consequently, photon antibunching is inconsistent with the classical electromagnetic theory of light. In contrast, the next section will demonstrate that quantum effects in single-molecule or single-atom light sources can naturally result in such an antibunching effect.

7.2   SINGLE-MOLECULE EMITTERS: PHOTON ANTIBUNCHING

The simplest model of single-molecule emission involves two energy levels, the ground state G and an excited state E. An experimenter can excite the molecule from G to E using light from a laser. This excitation process can often be modeled as a first-order transition with a rate coefficient k_{ex}, which depends on the light intensity. An excited molecule can then jump back into the G, with the excess energy emitted as a photon of light; this process is modeled as a first-order transition back to G, with a rate coefficient Γ. The overall kinetic scheme of this process is as follows:

$$G \underset{\Gamma}{\overset{k_{ex}}{\rightleftharpoons}} E.$$

   A key property of the above process is that, immediately after a photon has been emitted, the molecule is found in its ground state and, therefore, it cannot emit another photon. Thus successive photons "antibunch" or "repel" one another. To understand this property more quantitatively, consider the probability distribution $w_{lag}(t)$ of the lag time between two successive photons in this case. If $t = 0$ corresponds to the emission of the last photon, then $w_{lag}(t)dt$ is equal to the probability that the next photon is emitted between t and $t + dt$. The emission of the next photon proceeds via two steps. First, the molecule is excited, say, at some time t_1. Since the present kinetic

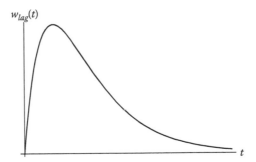

FIGURE 7.2 The lag time distribution for photons emitted by a single molecule is zero at short times (photon antibunching), displays a maximum at an intermediate time, and then eventually decays to zero.

scheme is formally equivalent to that discussed in Chapter 3, we can use Eq. 3.8 (with an appropriate notation change) to find the probability distribution of the waiting time t_1, which gives:

$$w_1(t_1) = k_{ex} e^{-k_{ex} t_1}.$$

Next, a transition back to G, accompanied by the emission of a photon, occurs. The probability distribution of the lag time $t_2 = t - t_1$ between excitation and emission is, again, given by an exponential:

$$w_2(t_2) = \Gamma e^{-\Gamma t_2}.$$

Since the times t_1 and t_2 are statistically independent their joint probability is the product $w_1 w_2$; to obtain the probability distribution of the lag time t we should integrate this product over all times subject to the constraint $t = t_1 + t_2$, i.e.,[2]

$$w_{lag}(t) = \int_0^t dt_1 w_1(t_1) w_2(t - t_1) = \frac{e^{-k_{ex} t} - e^{-\Gamma t}}{k_{ex}^{-1} - \Gamma^{-1}}. \qquad (7.14)$$

The shape of this function is illustrated in Figure 7.2. It is zero both at $t = 0$ and $t \to \infty$ and it reaches a maximum at an intermediate value of time t. This result is distinctly different from the Poisson case (Eq. 7.9), where $w_{lag}(t)$ decays monotonically. The existence of the dip exhibited by $w_{lag}(t)$ at short times t, usually referred to as photon antibunching, is a prominent signature of single-molecule emission. This property can be exploited by an experimenter to confirm that what she is looking at is just one molecule, as opposed to a collection of molecules aggregated next to one another.[3] Indeed, based on the discussion of Section 7.1, we expect the photon sequence from multiple molecules to universally lack the antibunching dip in its inter-photon lag time distribution.

---

[2] It is assumed here that $\Gamma \neq k_{ex}$. For $\Gamma = k_{ex}$ the integral below evaluates to $w_{lag}(t) = \Gamma^2 t \exp(-\Gamma t)$.
[3] Recall, from Chapter 2, that images of single molecules are blurred because the light wavelength is much greater than the typical molecular size. As a result, it is often impossible to tell directly whether one or more molecules are seen in an image.

Although photons emitted by a single molecule do not generally obey Poisson statistics, there are two limits where a Poisson process becomes a good approximation. In the first, the excitation rate is much higher than the emission rate, $k_{ex} \gg \Gamma$, and so excitation of the state E is almost instantaneous, as compared to the typical inter-photon lag time. Mathematically, the terms containing k_{ex} can be neglected in Eq. 7.14 except at very short times $t < k_{ex}^{-1} \ll \Gamma^{-1}$ and, at longer timescales, the lag time distribution is well approximated by the Poisson formula,

$$w_{lag}(t) \approx \Gamma \exp(-\Gamma t).$$

The second limit is the weak excitation case, $k_{ex} \ll \Gamma$, where, analogously to the previous case, the terms containing Γ can be neglected at all but very short times (shorter than Γ^{-1}) so that the resulting lag distribution can be approximated by

$$w_{lag}(t) \approx k_{ex} \exp(-k_{ex}t)$$

corresponding to a Poisson process with $\lambda \approx k_{ex}$. In this case, the average interphoton distance $\Delta t \approx k_{ex}^{-1}$ is controlled by the rate with which the molecule is excited into the state E, from which photons are quickly emitted.

Away from the above two limits, the mean photon count rate λ (i.e., the average number of photons emitted per unit time) differs from both k_{ex} and Γ. It can be determined by considering the probability for a photon to be emitted during an infinitesimal time interval dt, which, by definition, is equal to λdt. It is, on the other hand, equal to the conditional probability Γdt for a photon to be emitted provided that the molecule is in its light-emitting, excited state E, times the probability w_E to be in E, and so we have

$$\lambda = \Gamma w_E. \tag{7.15}$$

Assuming that the molecule is in its steady state (i.e., the laser has been continuously on for a while), the latter probability is constant and is easily estimated from the condition

$$dw_E/dt = k_{ex}w_G - \Gamma w_E = 0,$$

where $w_G = 1 - w_E$ is the probability to be in the ground state. This gives

$$w_E = \frac{k_{ex}}{k_{ex} + \Gamma}$$

and

$$\lambda = \frac{k_{ex}\Gamma}{k_{ex} + \Gamma}. \tag{7.16}$$

Exercise

Show that the same result can be obtained from

$$\lambda = \langle t_{lag} \rangle^{-1}$$

where

$$\langle t_{lag} \rangle = \int_0^\infty t w_{lag}(t)dt$$

is the mean interphoton lag time.

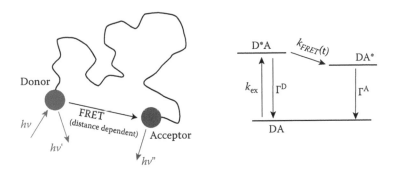

FIGURE 7.3 In a FRET (Fluorescence Resonance Energy Transfer) experiment the molecule, whose properties the experimentalist wishes to study (and represented here by a squiggly line), is labeled with two specially chosen fluorescent molecules called donor (D) and acceptor (A). Laser light is then tuned to promote the donor molecule to an excited state D*. The excited donor can then either reemit a photon thus going back to its ground state or transfer its excitation to the acceptor molecule via a quantum mechanism called FRET. Thus created excited acceptor A* then emits a photon of a lower frequency. Because the probability of excitation transfer is distance dependent, relative count rates of donor and acceptor photons provide information about the molecular configuration. The scheme on the right shows a simplified energy diagram for such a measurement.

7.3　MONITORING CONFORMATIONAL CHANGES WITH FLUORESCENCE RESONANCE ENERGY TRANSFER (FRET)

7.3.1　THE BASICS OF FRET

We will now introduce the simplest (and perhaps most common) experimental setup capable of monitoring molecular dynamics via light detection. This setup (illustrated in Figure 7.3) involves two single-molecule emitter probes, of the type considered in the previous section, attached to the molecule that one wishes to study. For reasons that will become clear below, the two probes will be referred to as the donor (D) and acceptor (A). In the absence of the acceptor, or when the acceptor is far away from the donor, the donor behaves as described in the previous section. That is, it can be excited to a higher-energy state D* by a laser that is tuned to a specific frequency ν at which the donor is likely to absorb light. This process is described by a rate coefficient k_{ex}. The excited donor can then reemit a photon with a frequency ν', a process described by a rate coefficient Γ^D. When the acceptor approaches the donor, however, a third process is possible, where the donor "donates" its energy and the acceptor "accepts" it through a quantum mechanical process known as the fluorescence resonance energy transfer, abbreviated as FRET. As a result, the acceptor is now in its excited state A* (while the donor is back to the ground state) and can emit a photon with a rate coefficient Γ^A. Note that the superscripts A or D are used here to label the two types of emitters rather than to represent exponentiation. Subscripts will be reserved to enumerate different molecular conformations introduced later in this chapter.

It may appear surprising that the FRET is a one-way process. Indeed, if the donor and acceptor energies are the same (as the word "resonance" in the name of the process implies), why would the reverse process, D* $\longleftarrow$ A*, not happen with the same probability? The answer is that the energy transfer to the acceptor instantaneously creates a non-thermal acceptor state (i.e., one that is not described by Boltzmann statistics), which then rapidly evolves into a state with a lower energy, through a relaxation process of the type considered in Chapter 4. A reverse process would require that the acceptor acquire, through a thermal fluctuation, additional energy to attain resonance with the donor, a process that tends to have low probability. The greatly simplified diagram shown in Fig.7.3 leaves these finer details of the process out: A more complete picture which accounts for the energy stored in the molecular vibrations and its flow during various transitions depicted in Fig.7.3 can be found in the literature [2].

Several other simplifying assumptions are made in the scheme of Fig.7.3. Firstly, while the acceptor is still in its excited state A*, there is nothing to prevent the donor from being excited again, resulting in a doubly excited state A*D*. It would be straightforward to incorporate this additional state into our scheme if desired. In practice, however, the probability of populating the doubly excited state is negligibly small unless the excitation rate k_{ex} is very high, much higher than the emission rate Γ^A, a case that will not be studied here. Likewise, the possibility of transitions to other molecular states that are not present in Fig.7.3 will be ignored. Finally, although the FRET probes are usually chosen so as to avoid direct excitation of the acceptor by the laser, some small probability of such a process commonly remains. Again, the direct excitation of the acceptor would be straightforward to include in our scheme but it will be ignored in order to keep the discussion as simple as possible.

The key property of the FRET system that makes it useful is that the rate coefficient for the energy transfer, k_{FRET}, is very sensitive to the donor-acceptor distance[4] and so any molecular transition that changes this distance manifests itself as a change in the emission pattern. If, for example, the donor and acceptor are far apart, energy transfer is unlikely and only donor photons are detected, but when a molecular transition brings the two probes together then acceptor emission becomes significant. To understand how the value of k_{FRET} influences this observable FRET signal, we start with the case where k_{FRET} takes on a constant value. What are the resulting count rates for donor (λ^D) and acceptor (λ^A) photons? These can be estimated similarly to Eq. 7.15:

$$\lambda^{D(A)} = \Gamma^{D(A)} w^{D(A)}, \tag{7.17}$$

where $w^{D(A)}$ is the steady-state probability of having the donor(acceptor) excited. Let $w^0 = 1 - w^D - w^A$ be the steady-state probability to be in the DA state where neither of the two probes is excited. Then w^D and w^A satisfy the balance equations:

$$dw^D/dt = k_{ex}w^0 - \Gamma^D w^D - k_{FRET}w^D = 0$$
$$dw^A/dt = k_{FRET}w^D - \Gamma^A w^A = 0.$$

[4] More specifically, k_{FRET} is inversely proportional to the sixth power of the distance between the donor and the acceptor.

From the first of these equations we find

$$w^D = \frac{k_{ex} w^0}{\Gamma^D + k_{FRET}}, \tag{7.18}$$

and from the second we have

$$w^A = \frac{k_{FRET} k_{ex} w^0}{(\Gamma^D + k_{FRET}) \Gamma^A}. \tag{7.19}$$

The donor and acceptor photon count rates then can be written as

$$\lambda^D = \Gamma^D w^D = \lambda(1 - \epsilon) \tag{7.20}$$

$$\lambda^A = \Gamma^A w^A = \lambda\epsilon, \tag{7.21}$$

where

$$\epsilon = \frac{k_{FRET}}{k_{FRET} + \Gamma^D} \tag{7.22}$$

and

$$\lambda = k_{ex} w^0. \tag{7.23}$$

The parameter ϵ is called the FRET efficiency: It is the probability that an excited donor will transfer its energy to the acceptor, eventually causing the emission of an acceptor photon, while $1 - \epsilon$ is the probability that the donor will emit a photon instead. Because, in our scheme, each excitation event (which occurs at a rate $k_{ex} w^0$) results in the emission of a photon, the parameter λ represents the total photon count rate. In particular, we have

$$\lambda^D + \lambda^A = \lambda. \tag{7.24}$$

In Section 7.2 we saw that in the weak excitation limit the photon antibunching effect can be neglected and the resulting emission is well approximated by Poisson statistics. This approximation will be adopted here to simplify further discussion; that is, we will be viewing both the donor and the acceptor as Poisson emitters producing photons at rates λ^D and λ^A, respectively. Moreover, the population w^0 can now be replaced by $w^0 = 1 - w^A - w^D \approx 1$, as it is readily seen from Eqs.7.18 and 7.19 that the probabilities for the donor and acceptor to be excited satisfy the inequalities $w^A, w^D \ll 1$ if $k_{ex} \ll \Gamma^A, \Gamma^D$. Therefore, in this limit, the total count rate $\lambda = k_{ex}$ is independent of the properties of the molecule that is being studied.

Now let us consider the more interesting case where intramolecular dynamics affects the distance between the donor and the acceptor. As a result, the FRET rate coefficient k_{FRET} and the FRET efficiency become functions of time. Within the assumptions made above, the total photoemission rate λ remains constant and insensitive to the dynamics (cf. Eq. 7.24). That is, an experiment that does not resolve photon colors carries no information about the molecular motion. However, the count rates for photons of each individual color, i.e., $\lambda^D(t) = \lambda[1 - \epsilon(t)]$ and $\lambda^A(t) = \lambda\epsilon(t)$, are now time-dependent. A commonly occurring scenario of this kind arises when the molecule jumps between two distinct conformations, which will be referred to as 1 and 2, with their interconversion being described by first-order kinetics,

$$1 \underset{k_0}{\overset{k_0}{\rightleftharpoons}} 2.$$

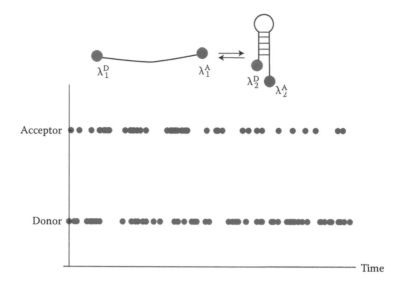

FIGURE 7.4 A computer-generated FRET signal from a system that undergoes jumps between two states (e.g., hairpin and linear DNA conformations shown at the top) with different distances between the donor and the acceptor. Blue and red circles indicate, respectively, donor and acceptor photons. The FRET efficiency in state 1 is $\epsilon_1 = 0.25$ so that 75% photons emitted in this state are from the donor. The FRET efficiency in state 2 is $\epsilon_2 = 0.75$ so 75% of photons arriving from this state are acceptor photons. Based on this information, can you infer how the state of the system evolved as a function of time?

Although the backward and forward interconversion rate coefficients are assumed to be equal to the same number k_0 in order to simplify the algebra, this assumption is not essential and can be easily lifted if desired. Figure 7.4 illustrates this scenario and shows computer-simulated streams of donor and acceptor photons. In this figure, the value of the FRET efficiency in state 1 was assumed to be $\epsilon_1 = 0.25$, which means that the majority of the photons emitted from this state are donor photons. In state 2, however, the FRET efficiency was taken to be $\epsilon_2 = 0.75$, resulting in predominantly acceptor photons.

The challenge is now to recover the underlying dynamics of the molecule (i.e., its state as a function of time) from the observed sequences of donor and acceptor photons exemplified by Figure 7.4. Naturally, when we see lots of donor photons we expect the system to be in state 1, while emission dominated by acceptor photons is indicative of state 2. Using this observation, the reader is encouraged to take a pencil and try to make a sketch of the molecule's state (1 or 2) as a function of time, while not looking at Figure 7.5. Now compare your sketch with the actual trajectory that has led to the photon sequences shown in Figure 7.4—this trajectory is shown in Figure 7.5 as a dashed line. I bet your answer is not the same! Keep in mind that your task was relatively easy because you knew in advance that the model contained only two states. But what if there were more states? Could you tell how many states you see by looking at Figure 7.4?

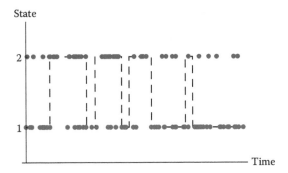

State

FIGURE 7.5 The same data as in the previous figure are now overlaid with the underlying trajectory of the system (i.e., the state of the system as a function of time) shown as a dashed line.

The above example illustrates some of the predicaments that single-molecule experimenters face. First, no one-to-one correspondence exists between the photon color and the corresponding state. That is, photons of both colors are emitted from both states, albeit in different proportions. Second, the lag times between the photons are finite and so nothing is known about the molecule during the dark times when no photons are detected. What can one do to overcome these difficulties?

7.3.2 BINNING

One way to estimate the instantaneous values of the donor and acceptor count rates (and thus infer the corresponding molecular state) is by counting the numbers $n^D(\Delta t)$ and $n^A(\Delta t)$ of, respectively, donor and acceptor photons collected over a certain time interval Δt. Specifically, we write:

$$\lambda^{D(A)} \approx n^{D(A)}(\Delta t)/\Delta t,$$

and the instantaneous FRET efficiency is then estimated as

$$\epsilon = \frac{\lambda^A}{\lambda^D + \lambda^A} = \frac{n^A}{n^D + n^A}.$$

By repeating this step for successive time bins defined by $t_i \leq t < t_{i+1} = t_i + \Delta t$, a discrete trajectory composed of the values of $\epsilon(t_1)$, $\epsilon(t_2)$, . . ., is obtained.

An example of a trajectory $\epsilon(t)$ thus estimated is shown in Fig. 7.6 (bottom). Unlike in a real experiment, this curve was produced using computer simulation data and so the actual trajectory of the two-state system that has led to the observed photon pattern was also known (shown in Fig. 7.6, top). The comparison of the two time dependences thus informs us about the accuracy of the binning method. We see that, despite resemblance between the two, the estimated time dependence of $\epsilon(t)$ is quite noisy. This noise is caused by fluctuations in the number of photons within each time bin around its expected value. According to Eq. 7.8, the magnitude of those fluctuations can be estimated as

$$\Delta n^{D(A)} \approx \sqrt{\lambda^{D(A)}\Delta t} \approx \sqrt{n^{D(A)}}$$

and so the relative error in measuring the value of $\lambda^{D(A)}$ is $\Delta n^{D(A)}/n^{D(A)} \approx 1/$ $\sqrt{\lambda^{D(A)}\Delta t}$. This error can be reduced by using a larger bin size Δt.[5] Unfortunately, this leads to poorer time resolution since any conformational changes occurring at time scales shorter than Δt can no longer be resolved. For the particular case of two-state dynamics considered here, the timescale to be concerned with is the typical dwell time in each state (i.e., k_0^{-1}) and so the bin size Δt must be much shorter than k_0^{-1} to capture essential features of the dynamics. In the example shown in Fig. 7.6, the bin size was chosen to be only an order of magnitude shorter than the timescale of interest, $\Delta t = 0.1k_0^{-1}$. Yet given the assumed total count rate of $\lambda = 300/k_0$, the relative error of measuring the count rate is $1/\sqrt{\lambda\Delta t} = 1/\sqrt{30} \approx 20\%$. Moreover, the actual error in the count rate for each type of photons (with the associated value of ϵ) is higher (because λ^A and λ^D are always lower than λ) and depends on the current state. To reduce the noise, say, by a factor of 3, the bin size would have to be increased by a factor of $3^2 = 9$, resulting in $\Delta t = 0.9/k_0$. But such a large bin size, comparable to the timescale of interest, would render the estimated $\epsilon(t)$ completely meaningless.

The above example shows that the competing requirements for the bin size to be small enough to provide sufficient time resolution yet large enough to suppress the noise put the experimenter between a rock and a hard place. In some cases, binning makes no sense at all. For example, when the same binning analysis was applied to Figure 7.4 (where the value of λ was about 30 times lower), the resulting $\epsilon(t)$ was dominated by the noise providing no useful information whatsoever (the data are not shown because they were not worth the space in this book).

What is even more disconcerting about Fig. 7.6 is that it fails to clearly show the two-state behavior of the underlying system. Try to pretend that you have not been told that the system had two states—how many states do you really see there? If unambiguous identification of the number of states is difficult for a simple two-state model with a known value of k_0, think how hard it would be in a real experiment, where the lack of prior knowledge about the system's characteristic timscales leaves the experimenter without any idea as to what an acceptable bin size should be!

In fact, the *apparent* number of states may depend on the bin size. Consider, for example, the case of a very large bin size, $\Delta t \gg k_0^{-1}, \lambda^{-1}$. During the time Δt, the molecule then undergoes numerous jumps between the states and emits a large number of photons (i.e., $n^D, n^A \gg 1$). The number of, say, acceptor photons detected within a bin can then be estimated as

$$n^A(\Delta t) = \Delta t(w_1\lambda_1^A + w_2\lambda_2^A) = \lambda\Delta t(w_1\epsilon_1 + w_2\epsilon_2).$$

Here, the subscript (1 or 2) is used to denote the state the system is in while the superscript (D or A), as before, indicates whether the associated photon originates

---

[5] Experimentally, this error can also be reduced by increasing the laser power and, therefore, the value of λ. Although a useful approach in some cases, it usually shortens the lifespan of the fluorescent donor and acceptor, which tend to "photobleach," i.e., undergo an irreversible photochemical transformation after which they no longer emit light.

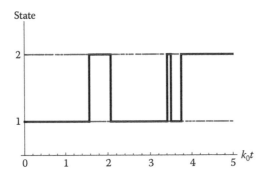

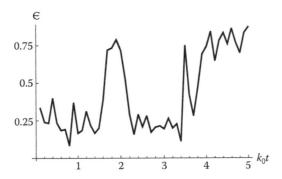

FIGURE 7.6 Top: simulated photon arrival times (blue: donor, red: acceptor) overlaid with the underlying time evolution of a molecular state (dashed line). Bottom: The time evolution of the system was estimated by averaging the FRET efficiency ϵ over time bins of length $\Delta t = 0.1/k_0$, where k_0 is the rate coefficient for jumps between states 1 and 2.

from the donor or the acceptor. For example, w_2 is the equilibrium probability to be in state 2 while λ_1^D is the count rate from the donor when the system is in state 1. A similar expression can be written for the number of donor photons $n^D(\Delta t)$ by simply replacing $\epsilon_{1,2}$ by $1 - \epsilon_{1,2}$. The FRET efficiency observed within this bin is then

$$\frac{n^A(\Delta t)}{n^A(\Delta t) + n^D(\Delta t)} = w_1\epsilon_1 + w_2\epsilon_2 = \frac{1}{2}(\epsilon_1 + \epsilon_2).$$

Because of the large bin size and, consequently, large number of photons, fluctuations in the number of photons and, therefore, of the measured FRET efficiency can be neglected. The system therefore appears to be in a single state, with an apparent FRET efficiency equal to the equilibrium average of ϵ over the two states. Any information about the molecule's dynamics is lost.

To recover dynamical information we must decrease the bin size. As we do so $\epsilon(t)$ becomes time-dependent both because the state of the molecule changes and because the number of photons fluctuates within a bin. In the most favorable scenario, there may be an optimal range of values of Δt such that the relative error in the count rate within a bin is small while the bin is small enough to resolve jumps

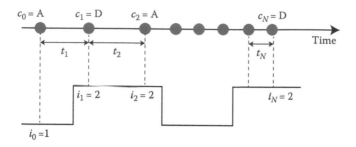

FIGURE 7.7 A sequence of $N + 1$ photons is characterized by the lag times $\{t_1, t_2, \ldots, t_N\}$ between consecutive photons and by the photon colors $\{c_0, c_1, c_2, \ldots, c_N\}$ $(c_i = D, A)$. At the moment the i-th photon is emitted, the system is in (generally unknown) state s_i. To estimate the probability of each photon sequence, one needs to perform an average over all possible state sequences $\{i_0, i_1, \ldots, i_N\}$.

between individual states. Unfortunately, this is often not the case: More commonly, an estimate of $\epsilon(t)$ obtained via binning is strongly dependent on the bin size and the true dynamics of the molecule is entangled with the noise. When more than two states are implicated, the analysis of the data becomes even more daunting because discontinuous hopping among multiple states is increasingly harder to distinguish from the noisy $\epsilon(t)$ produced by binning and because it may be hard (or impossible) to find a bin size that is simultaneously optimal with respect to all the relevant hopping timescales.

7.3.3 INTERPHOTON LAG TIMES AND INTENSITY AUTOCORRELATION FUNCTIONS

Rather than attempting to learn precisely what molecular trajectory has led to the observed photon emission pattern, one may try constructing a model that fits certain statistical properties of the signal. This approach is particularly appealing because molecular trajectories are stochastic and therefore irreproducible. The specific trajectory recorded in any particular experiment is, therefore, of little interest *per se*. Rather, any meaningful trajectory analysis should be formulated in statistical terms.

One statistical property that can be easily extracted from the raw experimental data (i.e., recorded photon arrival times illustrated in Fig.7.7) is the probability distribution of the inter-photon lag times. One can, for example, identify all the instances where an acceptor photon is followed by another acceptor photon. By generating a histogram of the lag times between two successive acceptor photons for all those instances, the lag time distribution $w^{AA}(t)$ is estimated. For example, if the photons in the sequence shown in Fig.7.7 are numbered from zero to $N = 7$, then the times between the second and the third and between the fifth and the sixth photons will contribute into $w^{AA}(t)$. In a similar manner, we can estimate the distribution $w^{DA}(t)$ of the lag times for an acceptor photon followed by a donor one. Note that, as before, time evolves from right to left in all our mathematical expressions and so the superscript DA means that the acceptor photon precedes the donor one. In what follows, we will estimate

$w^{DA}(t)$ and $w^{AA}(t)$ for our two-state system (and, more generally, for any discrete system that can be described by a master equation).

Exercise

Can $w^{DA}(t)$ be different from $w^{AD}(t)$? Can $w^{DD}(t)$ be different from $w^{AA}(t)$?

Consider the following sequence of events: At $t = 0$ an acceptor photon is emitted, while the molecule is in state i. At time t a donor photon is emitted while the molecule is in state j. No photons are emitted between 0 and t. What is the probability of this sequence? The probability of the emission of an acceptor photon is proportional to λ_i^A.[6] The probability that a donor photon is emitted at time t is, likewise, proportional to λ_j^D. As found before, the probability that no other photon (regardless of its color) intervenes between 0 and t is $e^{-\lambda t}$. Therefore, $w^{DA}(t)$ is proportional to

$$e^{-\lambda t}\lambda_j^D\lambda_i^A$$

averaged over the possible states i and j. The weight, with which each such pair of states enters the average, is the joint probability that the system is in state i at time 0 and in state j at time t and can be written as $T_{ji}(t)w_i$, where w_i is the equilibrium probability of being in state i and $T_{ji}(t)$ is the probability of being at j at time t conditional upon having started in i at $t = 0$ (see Chapter 4). Putting it all together, and summing over all possible states we obtain:

$$w^{DA}(t) = N^{DA}\sum_{i,j}\lambda_j^D e^{-\lambda t}T_{ji}(t)\lambda_i^A w_i, \qquad (7.25)$$

where N^{DA} is a normalization factor that ensures that the integral of $w^{DA}(t)$ over t is equal to 1. Eq. 7.25 is rather general and applies to any multi-state process obeying first order kinetics. Moreover, for any such process, the conditional probabilities, $T_{ji}(t)$, can be evaluated using matrix algebra as described in Chapter 4. Because in the present case we only have two states, $T_{ji}(t)$ can be evaluated in a simpler way. For example, $T_{11}(t)$ and $T_{21}(t)$ are, respectively, the probabilities to find the molecule in states 1 and 2 provided that it started in state 1 at $t = 0$. Consequently, they are the solutions of the system's kinetic equations

$$dT_{11}(t)/dt = -dT_{12}(t)/dt = -k_0 T_{11}(t) + k_0 T_{21}(t)$$

with the initial conditions $T_{11}(0) = 1$ and $T_{21}(0) = 0$. Since the two probabilities must add up to one, we further write

$$dT_{11}(t)/dt = -k_0 T_{11}(t) + k_0[1 - T_{11}(t)] = k_0[1 - 2T_{11}(t)],$$

---

[6] The probability to have a photon at exactly the time $t = 0$ is obviously zero but the probability of having this photon within an infinitesimal time interval dt is $\lambda_i^A dt$. The factors such as dt are ignored in this discussion, but the resulting probability distribution will be properly normalized at the end of the derivation.

which gives

$$T_{11}(t) = (1 + e^{-2k_0 t})/2,$$

and

$$T_{21}(t) = 1 - T_{11}(t) = (1 - e^{-2k_0 t})/2.$$

Similarly, starting from state 2, we obtain

$$T_{12}(t) = (1 - e^{-2k_0 t})/2$$

and

$$T_{22}(t) = (1 + e^{-2k_0 t})/2.$$

Substituting these into Eq. 7.25, after some algebra we finally find:

$$w^{DA}(t) = N^{DA} e^{-\lambda t}[(\lambda_1^D + \lambda_2^D)(\lambda_1^A + \lambda_2^A) + e^{-2k_0 t}(\lambda_1^D - \lambda_2^D)(\lambda_1^A - \lambda_2^A)]. \quad (7.26)$$

The normalization factor N^{DA} can be easily found, if desired, from the condition $\int_0^\infty dt\, w^{DA}(t) = 1$. The explicit expression for this factor is unimportant for our discussion.

As discussed before, when emission properties are independent of the state (or when there is only one state), photoemission is a Poisson process. Indeed, if $\lambda_1^D = \lambda_2^D$ and $\lambda_1^A = \lambda_2^A$, the second term in Eq. 7.26 disappears and it becomes a single exponential, $w^{DA}(t) \propto e^{-\lambda t}$. In contrast, when the donor and acceptor count rates change upon transitions, a second exponential, dependent on the transition timescale k_0, appears in Eq. 7.26. Since we have

$$(\lambda_1^D - \lambda_2^D)(\lambda_1^A - \lambda_2^A) = [\lambda(1 - \epsilon_1) - \lambda(1 - \epsilon_2)](\lambda\epsilon_1 - \lambda\epsilon_2) = -\lambda^2(\epsilon_1 - \epsilon_2)^2,$$

this term is always negative, thereby suppressing the likelihood that a photon of one color is quickly followed by a photon of the other color. Note, however, that if many photons are emitted during a typical time the molecule dwells in one state (i.e., $\lambda \gg k_0$), then $w^{DA}(t)$ becomes virtually indistinguishable from a single exponential ($\propto e^{-\lambda t}$). Indeed, the presence of the term $e^{-\lambda t}$ ensures that $w^{DA}(t)$ will essentially decay to zero on a timescale comparable to $1/\lambda$, during which the exponential $e^{-2k_0 t}$ remains virtually constant. A better way to highlight the effect of molecular dynamics on the photon statistics is to consider the quantity

$$C^{DA}(t) = w^{DA}(t)e^{\lambda t},$$

which evolves on the timescale of the molecule's dynamics (i.e., k_0^{-1}). C^{DA} can be obtained from Eq. 7.25 by eliminating the condition that no photon is emitted during the time interval t. In other words, it reports on the statistics of a lag between donor and acceptor photons regardless of how many other photons intervene in between. To within a proportionality coefficient, $C^{DA}(t)$ is the correlation function of the photon count rate:

$$\langle \lambda^D(t)\lambda^A(0)\rangle = \sum_{i,j} \lambda_j^D T_{ji}(t)\lambda_i^A w_i = (1/N^{DA})C^{DA}(t).$$

The lag time distribution for an acceptor (or donor) photon followed by a photon of the same color is given by an expression analogous to Eq. 7.26,

$$w^{AA}(t) = N^{AA}e^{-\lambda t}[(\lambda_1^A + \lambda_2^A)(\lambda_1^A + \lambda_2^A) + e^{-2k_0 t}(\lambda_1^A - \lambda_2^A)^2], \qquad (7.27)$$

while the corresponding function

$$C^{AA}(t) = w^{AA}(t)e^{\lambda t}$$

describes, to within a constant factor, the autocorrelation function of the acceptor count rate $\langle \lambda^A(t)\lambda^A(0) \rangle$. Note that, unlike in w^{DA} and C^{DA}, the term resulting from inter-state hopping (i.e., the one proportional to $e^{-2k_0 t}$) is always positive. This bunching of the photons of the same color is readily observable in Fig. 7.4: The photon color tends to remain the same until the molecule leaves the state it is in.

7.3.4 THE MAXIMUM LIKELIHOOD APPROACH

To implement the correlation function or lag time analysis in practice, the functions $w^{AA}(t)$, $C^{AA}(t)$, etc. must be measured. The model parameters (e.g., the hopping rate k_0 and the count rates $\lambda_{1(2)}^{D(A)}$) must then be adjusted to ensure that the expressions Eq. 7.27, Eq. 7.26 (or equivalent expressions for $C^{AA}(t)$ or $C^{DA}(t)$) fit the experimental data as well as possible. If the resulting fit is still poor, a different model (for example, one including more than two states) would be the next candidate to analyze, and so forth. This venerable approach to data analysis often shines when relative simplicity of the underlying dynamics or physical insight (or both) allows the experimenter to come up with an adequate model or when the information sought is insensitive to the model. FRET studies of protein dynamics in the unfolded state, for example, provide an example of such a success story (see, e.g., [4]) because such dynamics can be interpreted in terms of simple, continuous models dependent on few parameters.

On a fundamental level, however, this approach does not appear entirely satisfying. It seems to throw information away by focusing on just one specific property such as pairwise correlation between successive photons. For processes of sufficient complexity this strategy could be about as useful as trying to learn about Shakespeare's writing style by counting how often his plays contain a "b" followed by an "e." A further complication comes from the finite duration of single-molecular trajectories, which terminate when a fluorescent dye undergoes the irreversible photobleaching process. Given limited amount of data, correlation functions and lag time distributions estimated from individual trajectories are often too noisy; although the noise can be further reduced by averaging over multiple molecules, such averaging may defeat the purpose of a single-molecule study. On the other hand, despite its short duration a single photon sequence may contain additional information that could improve our chances to infer an adequate model. Is there a systematic way to utilize *all* of such information?

Methods employed by single-molecule researchers to accomplish this task are often borrowed from other areas of science and engineering. Before discussing those, let us reflect on some of the common difficulties encountered in single-molecule data analysis. First, random nature of photoemission results in data that are inherently noisy (cf. Fig. 7.6). While noise can be partially removed via binning or through averaging inherent in the computation of correlation functions, this noise removal also results in loss of information. Second, the underlying dynamics that has led to the observed photon sequences is generally unknown; for example, even a clear observation of two distinct states in the FRET signal does not necessarily mean that a two-state model of reaction kinetics is sufficient because other "dark" states may affect the dynamics while contributing no observable photons. Thus the right model must be guessed or deduced from the signal (more on this in the next section). In dealing with these problems, physicists and chemists are not alone. The problem of deciphering noisy data is ubiquitous in many areas that affect our daily lives. Your cell phone, for example, may have speech recognition software, which converts your voice (a noisy acoustic signal) into discrete series of letters and words. Economic forecasting is based on noisy data (e.g., fluctuations of price), from which it attempts to deduce models allowing prediction of the future. While numerous textbooks have been written on these and related topics, the scope of the present discussion will be limited to one particularly useful approach known as the maximum likelihood method. Starting with very simple examples, we will then work toward progressively more complicated scenarios encountered in FRET studies.

Suppose a particular observation has revealed $N + 1$ photons separated by lag times $t_1, t_2, \ldots, t_N$ (see Fig. 7.7). Let us first ignore the photon color and attempt to construct a "model" of the emission process that has resulted in this sequence. We could surmise that what we see is, in fact, a Poisson process, with a yet unknown count rate λ. If so, the probability density of the lag time t_i is given by Eq. 7.9. Since the Poisson process has the Markov property (no memory), the lag times are statistically independent and so the joint probability density of all the lag times is simply the product

$$\prod_{i=1}^{N} w_{lag}(t_i) = \lambda^N \exp\left[-\lambda \sum_{i=1}^{N} t_i\right].$$

Crudely speaking, this expression tells us how likely a particular sequence of lag times is given a specified value of λ. The maximum likelihood approach turns this assertion on its head and interprets this relationship as the likelihood of a specified value of the count rate λ given the observation of a particular sequence of lag times. That is, λ is now viewed as a variable in this expression. We call it the likelihood function and write:

$$L(\lambda|t_1, t_2, \ldots, t_N) = \prod_{i=1}^{N} w_{lag}(t_i) = \lambda^N \exp\left[-\lambda \sum_{i=1}^{N} t_i\right]. \tag{7.28}$$

It is also convenient to introduce the log likelihood function, which is just the natural logarithm of L:

$$\ln L(\lambda|t_1, t_2, \ldots, t_N) = N \ln \lambda - \lambda \sum_{i=1}^{N} t_i. \tag{7.29}$$

Notice that taking the log turns the product of statistically independent lag time distributions into a sum. In general, if several independent measurements are made, the log likelihood function is the sum of individual log likelihood functions.

We now determine the most likely Poisson model of the data by maximizing the likelihood function (or, equivalently, its logarithm) with respect to the model parameter λ. Taking derivative with respect to λ and setting it to zero, we obtain:

$$d \ln L(\lambda|t_1, t_2, \ldots, t_N)/d\lambda = N/\lambda - \sum_{i=1}^{N} t_i = 0$$

and

$$\lambda = N \bigg/ \sum_{i=1}^{N} t_i. \tag{7.30}$$

This result is not surprising: our estimate for λ is simply the total number of photons divided by the total observation time (equal to the sum of the lag times), exactly as expected from the definition of the mean count rate.

It should be emphasized that the maximum likelihood approach does not tell us whether the data were actually generated by a Poisson process. Indeed, given a finite set of lag times, we cannot possibly know, with certainty, if they resulted from a Poisson or some other process. Moreover, Eq. 7.30 provides an estimate of the count rate for the Poisson model regardless of whether or not this model is the right one. The likelihood function L, however, provides a measure of the likelihood that the observed data have resulted from a Poisson (or any other) process. If we suspect the process in question is not a Poisson process, we could try different models. The maximum likelihood method then allows one to compare the performance of different models via a comparison of their likelihood functions.

Exercise

Assuming that the lag times $t_1, t_2, \ldots, t_N$ obey a Gaussian distribution, i.e.,

$$w_{lag}(t) = \exp\left[-\frac{(t - t_m)^2}{2\sigma^2}\right] \bigg/ \sqrt{2\pi\sigma^2},$$

derive the maximum likelihood estimate for the mean t_m and the variance σ of this distribution.

The above analysis of colorless photon sequences can also be generalized to account for photon color. Suppose that, of the total of the $N + 1$ photons, N_D are donor photons and N_A acceptor photons. The simplest model of this process is the one where the FRET efficiency ϵ is constant. That is, the probabilities of finding that a given photon is emitted by the acceptor and donor are constant and equal, respectively, to ϵ and $1 - \epsilon$. The probability of observing N_D donor photons and N_A acceptor photons is then given by

$$C\epsilon^{N_A}(1 - \epsilon)^{N_D},$$

where C is the number of different photon sequences with N_A acceptor photons and N_D donor photons. Since C is independent of the model parameters ϵ and λ, its exact value is unimportant. Assuming that the color of a photon is statistically independent of the photon arrival times, we can now write the likelihood function, given the observed interphoton lag times and given the numbers of donor and acceptor photons, as the product:

$$L(\lambda, \epsilon | N_A, N_D, t_1, t_2, \ldots, t_N) = C\epsilon^{N_A}(1 - \epsilon)^{N_D}\lambda^N \exp\left[-\lambda \sum_{i-1}^{N} t_i\right]. \quad (7.31)$$

To find the optimal model parameters, we maximize this with respect to ϵ and λ. Since Eq. 7.31 has the form of a product of a function of ϵ and a function of λ, we can maximize each term in this product separately. Equivalently, the log likelihood function is the sum of two independent terms, one a function of λ and the other a function of ϵ. Finding the maximum with respect to λ results in the previously estimated count rate, Eq. 7.30. To maximize the likelihood with respect to ϵ, we write

$$d \ln[C\epsilon^{N_A}(1 - \epsilon)^{N_D}]/d\epsilon = \frac{N_A}{\epsilon} - \frac{N_D}{1 - \epsilon} = 0,$$

which gives

$$\epsilon = \frac{N_A}{N_D + N_A}.$$

This result shows that the probability that a photon is emitted by the acceptor can be estimated as the fraction of acceptor photons in the sequence. Again, no surprises here: We could have arrived at the same result without resorting to fancy statistical methods. The simplicity of this result, however, stems from the simplicity of the model used, where the molecule was assumed to remain in the same state.

We will now extend these ideas to construct a likelihood function for models that account for the change in the molecule's state (and, consequently, in the instantaneous FRET efficiency) as a function of time. The experimentally recorded sequence of photons illustrated Fig. 7.7 is now assumed to originate from a molecule that undergoes hopping among a set of discrete states. Each state i is characterized by its own value of the FRET efficiency ϵ_i and, correspondingly, by the donor ($\lambda_i^D = \lambda[1 - \epsilon_i]$) and acceptor ($\lambda_i^A = \lambda\epsilon_i$) photon count rates. The experimentally accessible information consists of the color of each photon, which is recorded in a string $c_0, c_1, c_3, \ldots, c_N$ (where $c_m =$ D or A), and the photon arrival times, or, equivalently, the lag times $t_1, t_2, \ldots, t_N$ between successive photons. Let $i_0, i_1, \ldots, i_N$ be the sequence of states occupied by the molecule at the moments when the respective photons are detected, as illustrated in Fig.7.7 for the case of two states (i.e., $i_m = 1$ or 2). Unlike the photon arrival times and their colors, this sequence of molecule's states is unknown to the observer.

Proceeding in a manner similar to the above two examples, we write that the probability of measuring a specific photon sequence is proportional to

$$\lambda_{i_N}^{c_N} e^{-\lambda t_N} \ldots e^{-\lambda t_2} \lambda_{i_1}^{c_1} e^{-\lambda t_1} \lambda_{i_0}^{c_0} = \lambda_{i_0}^{c_0} e^{-\lambda \sum_{m=1}^{N} t_m} \prod_{m=1}^{N} \lambda_{i_m}^{c_m}. \quad (7.32)$$

A generalization of Eq. 7.31, this formula states that the probability of emitting a photon of color c_m while in state i_m is proportional to $\lambda_{i_m}^{c_m}$ while the probability of having no photons emitted during the time interval t_m in between the $(m - 1)$-th and

m-th photons is $\exp(-\lambda t_m)$; Eq. 7.32 is obtained by taking the product of all of these probabilities. Viewed as a function of the sequence of molecular states, $i_1, i_2, \ldots, i_N$, Eq. 7.32 could be interpreted as a likelihood function. That is, maximizing this function with respect to $i_1, i_2, \ldots, i_N$ will produce the most probable sequence of events that has led to the observed pattern of photons. It easy to see that the outcome of this procedure will produce the sequence that only contains two states. Specifically, if the m-th photon has been emitted by the donor (acceptor), then the corresponding state i_m will be the one that maximizes the donor (acceptor) count rate. While such a prediction could be deemed reasonable for a two-state system, it clearly fails when more than two states are involved.

The problem with the above reasoning is that it incorporates no advance knowledge about the properties of the state sequence or the molecular mechanism behind it. Crudely speaking, it deems all possible state sequences equally likely. Taras Plakhotnik illustrates this point in [3] using the joke about a person who, when asked about the probability of being eaten by a dinosaur on the streets of the New York City, replies that this probability is exactly 50%, since there are only two possible outcomes, that one either gets eaten or not. This absurd answer does not take into account our knowledge about dinosaurs, particularly the fact that they have been extinct a long time ago. Surely, an insurance company using this kind of logic to estimate insurance costs will go bankrupt in no time!

To avoid this pitfall, we need to incorporate our knowledge, or educated guess, regarding the process through which the system hops among its states. This knowledge could, for example, include the fact that such hops cannot be too rare or too frequent, or the constraint that hops between only certain pairs of states are possible. This knowledge further consists of the model we use to describe the data as well as of physical constraints on its parameters. Provided that the hopping dynamics is described by a master equation, Eq. 4.20, the probability of any specific sequence of states, $i_1, i_2, \ldots, i_N$, can be written as the initial (equilibrium) probability to be in state i_0 times the product of conditional probabilities describing this sequence of states.

$$T_{i_N i_{N-1}}(t_N) \ldots T_{i_2 i_1}(t_2) T_{i_1 i_0}(t_1) w_{i_0}.$$

Our likelihood function, weighted with this sequence probability, now becomes

$$e^{-\lambda \sum_{m=1}^{N} t_m} \lambda_{i_N}^{c_N} T_{i_N i_{N-1}}(t_N) \ldots \lambda_{i_2}^{c_2} T_{i_2 i_1}(t_2) \lambda_{i_1}^{c_1} T_{i_1 i_0}(t_1) \lambda_{i_0}^{c_0} w_{i_0}. \qquad (7.33)$$

If the rate coefficients for the transitions between each pair of states (cf. Eq. 4.20) are known then the conditional probabilities, as well as the equilibrium populations, can be computed. Technical details of such a calculation are given in Section 4.5. Then, the above likelihood function can be maximized with respect to the sequence $i_1, i_2, \ldots, i_N$, resulting in a "most likely" trajectory consistent with the underlying dynamic model. It is, however, more common in single-molecule studies that the underlying model is the very object of the data analysis. At the same time, recovering the specific molecular trajectory that has led to the observed photon sequence is not important since this trajectory could not be reproduced in another experiment anyway. It then makes sense to regard the above likelihood function as a function of the model parameters (i.e., the transition rate coefficients $k_{i \to j}$ in Eq. 4.20). Furthermore, since the sequence $i_1, i_2, \ldots, i_N$ itself is unknown, weighted average over all possible sequences should be taken. A a result, we arrive at the following likelihood function:[7]

$$L(\lambda, \{k_{i \to j}\}, \{\lambda_i^{D,A}\} | \{t_i\}, \{c_i\})$$

$$= \sum_{i_0, \ldots, i_N} e^{-\lambda \sum_{m=1}^{N} t_m} \lambda_{i_N}^{c_N} T_{i_N i_{N-1}}(t_N) \ldots \lambda_{i_2}^{c_2} T_{i_2 i_1}(t_2) \lambda_{i_1}^{c_1} T_{i_1 i_0}(t_1) \lambda_{i_0}^{c_0} w_{i_0}. \quad (7.34)$$

An optimal model can now be estimated from the photon colors and arrival times by maximizing this likelihood function with respect to the model parameters (e.g., the rate coefficients and the emission rates in each state). Technical details of how this can be accomplished are given in the literature, see, e.g., [5].

7.4 RANDOM THOUGHTS ON COMPUTER-AIDED APPROACHES TO DISCOVERING SINGLE-MOLECULE DYNAMICS

Although our survey of single-molecule data analysis methods is far from complete, two distinct strategies can be identified. I will illustrate them using Figure 7.6 as an example of "experimental data." An experimenter adopting the first, "old-fashioned" approach combines physical intuition and/or prior knowledge about the system with observations to come up with a (preferably simple) model, whose parameters are then derived by fitting information extracted from the data (e.g., correlation functions, lag time distributions, etc.) to the predictions of the model. For example, although the underlying two-state behavior is not immediately apparent in Fig.7.6, the researcher may take advantage of her or his knowledge that the molecule in question has two chemically distinct conformations to propose a two-state model governed by first-order kinetics. The unknown parameters, i.e., the transition rate coefficients between the two conformations, are then estimated by fitting the data, using any of the methods described above. If the fit is poor or if some salient features of the data remain unexplained by the model, a more sophisticated model is attempted. For example, a two-state model predicts that the autocorrelation function of the FRET efficiency displays a single-exponential decay,[8]

$$\langle \epsilon(0) \epsilon(t) \rangle = a + b \exp\left[-t / (k_{1 \to 2} + k_{2 \to 1}) \right].$$

If, in contrast, the actual time dependence of this function happens to be multi-exponential, then more than two states have to be included in the model.

The second strategy adopts a more open-minded view and, in its most extreme form, attempts to uncover the information contained in the data without having any

---

[7] Notice that, when deriving Eq. 7.25, we have effectively computed this likelihood function for the case of a two-photon sequence (acceptor photon followed by a donor one). Eq. 7.34 generalizes that result to multi-photon sequences.

[8] Single-exponential behavior can be demonstrated using the matrix approach to the master equation described in Section 4.5. Indeed, the matrix $\mathbf{K}$ discussed there has only one nonzero eigenvalue, equal to $k_{1 \to 2} + k_{2 \to 1}$, and so all the transition probabilities T_{ij} display single-exponential behavior. This behavior is also apparent from the structure of the matrix of conditional probabilities T_{ij} derived for a two-state system in the preceding discussion because the only time dependence of this matrix comes from the exponential $\exp(-2k_0 t)$.

preconceived notion regarding the underlying mechanism. I will call it the automated, or machine learning, approach because the deciphering of the data is typically accomplished through the application of a computer algorithm. The challenge faced by machine learning is that of separating information from noise: Indeed, in the absence of noise seen in Fig. 7.6, there would be a one-to-one correspondence between the observed value of $\epsilon(t)$ and the state of the molecule (i.e., 1 or 2). Thus the underlying molecular trajectory would be satisfactorily uncovered.[9] If we write

$$\epsilon(t) = \epsilon_{true}(t) + \epsilon_{noise}(t), \tag{7.35}$$

where the two terms are the "true" signal and the noise, then the challenge is to extract the first term alone. Algorithms that achieve such denoising of signals are ubiquitous in science and engineering. Among their many uses is, for example, restoration of old photographic images and sound recordings. A related family of algorithms deals with "recognition of patterns" in sound or electromagnetic signals and includes, for example, the already mentioned speech recognition software. Are such algorithms useful in the context of single-molecule data analysis and could they outperform the old-fashioned strategies? The following comments represent the author's subjective opinion that may not be shared by all researchers in the field.

Let us consider the signal of Fig.7.6 and ask whether a clever computer code could recognize the underlying two-state pattern that would be seen in the absence of noise. This objective is, in many respects, analogous to the task performed by a speech recognition program. Do such programs perform better than humans? The speech recognition software on my brand new cell phone seems to butcher every other word I say. Perhaps my Russian accent plays a role here, yet when it comes to discerning other people's English, whether they are native speakers or not, I am confident I can beat my phone anytime. It has been long recognized that our brain has an amazing ability to recognize faces, speech, and other audio or visual patterns. Despite many efforts to improve machine learning algorithms, they remain inferior to humans. If so, how can we expect that a computer code will outperform a smart PhD student in recognizing the underlying two-state pattern seen in Fig.7.6?

It should be further noted that the goals of automated algorithms employed for single-molecule data analysis and for the more common tasks of pattern recognition are different. The purpose of phone speech recognition software, for example, is to automate certain routine tasks (e.g., to replace typing with voice commands while driving a car) rather than to outperform the phone owner. Likewise, face recognition software can spare an individual's effort to sieve through multiple photographic images even if it is inferior to the face recognition skill of a human. In contrast, automated single-molecule data analysis is aimed at improving upon the experimenter's ability to decipher the data. Consequently, while widely useful in many other contexts, machine learning methods may still be no match to the skill and the insight of a scientist.

---

[9] It should also be mentioned that, while the noise observed in Fig. 7.6 is entirely caused by the fluctuations in the number of photons within a time bin, other noise sources are usually present in real experiments. For example, "background" photons that have been emitted by sources other than the donor or acceptor may contribute to the observed signal.

Another notable complication inherent to single-molecule data analysis is that, contrary to what Eq. 7.35 appears to imply, the contribution of the noise to the overall signal is not independent of the signal itself. For example, the magnitude of the typical fluctuations in the number n of photons detected within a single time bin itself depends on n (cf. Eq. 7.8). As a result, the two terms in Eq. 7.35 are not independent of each other. This significantly complicates the denoising problem, as most standard algorithms assume additivity of the noise and the signal.

One of the further dangers of trying to uncover a hidden pattern from a noisy signal is to look for something that is simply not there. Recent efforts to denoise a 19th century piano recording of Johannes Brahms provide an illustration of this point. As the original recording is so poor that it is barely possible to even recognize the piece Brahms is playing (which happens to be his first Hungarian dance), this recording has provided a worthy challenge to the researchers working on denoising algorithms. Despite many attempts made to improve the quality of the original recording, however, we are presently not any closer to learning what Brahms' piano playing sounded like than we were with the original recording.[10] The denoised versions of the recording do feature less noise, as purported. The problem is not with the noise but with the rudimentary signal itself, which was recorded by one of the early models of the Edison wax cylinder. Single-molecule experiments often share this difficulty: Just like the richness of the sound produced by a great musician could not be faithfully recorded by a groove on a surface of a wax cylinder, the complexity of the motion performed by polyatomic molecules is unlikely to be fully reflected in a one-dimensional stream of photons. Development of more sophisticated methods such as multi-color FRET (which reports on several intramolecular distances at once) should increase information content of single-molecule measurements. Yet analysis of single-molecule data will likely remain to rely on physical intuition and insight. This is not to say that various machine learning algorithms are not useful. They are valuable tools but they make no miracles happen.

Finally, let us point out yet another limitation of single-molecule data analysis: Even under ideal conditions (negligible noise) there is no guarantee that there is a single, unique model consistent with the data. In fact, as discussed by O. Flomenbom and R.J.Silbey in [3], a multitude of models are equally consistent with the data and so the experimenter can never prove that his or her favorite model is the right one. Curiously, this is not necessarily a bad thing. For example, the measurement of protein folding transit times reported in [6] employed maximum likelihood analysis of photon arrival times (developed in [5]) as follows. If protein folding is viewed as a two-state process, then, by definition, the transition between the two states is instantaneous and the transit time is undefined. If, however, a third, intermediate state is introduced into the model, then the time the protein dwells in this state en route from the initial to the final state is the transit time. Using the maximum likelihood approach, the rate coefficients of the three-state model were determined and thus the transit time was estimated. But, of course, there is no reason to assume a single intermediate state since such a state is not directly observed. There could be two states, ten states, or a continuum of states! Luckily, fitting the data using various multi-state models gives

---

[10] This is my subjective opinion. Other listeners may disagree.

essentially the same estimate of the transit time. It is therefore the robustness of the findings reported in [6] and their insensitivity to the specifics of the model used that makes the results so compelling.

REFERENCES

1. Mark Fox, *Quantum Optics: An Introduction*, Oxford University Press, 2006.
2. David L. Andrews and Andrey A. Demidov (Editors), *Resonance Energy Transfer*, Wiley, 1999.
3. Eli Barkai, Frank Brown, Michel Orrit, and Haw Yang (Editors), *Theory and Evaluation of Single-Molecule Signals*, World Scientific, 2008.
4. D. Nettels, I.V. Gopich, A. Hoffman, and B. Schuler, "Ultrafast dynamics of protein collapse from single-molecule photon statistics", *Proc Natl Acad Sci USA*, v. 104, p. 2655, 2007.
5. Irina V. Gopich and Attila Szabo, "Decoding the pattern of photon colors in single-molecule FRET", *J. Phys. Chem. B*, vol. 113, 10965-10973, 2009.
6. Hoi Sung Chung, Kevin McHale, John M. Louis, and William A. Eaton, "Single-molecule fluorescence experiments determine protein folding transition path times," *Science*, vol. 335, pp. 981-984, 2012.

8 Single-Molecule Mechanics

When you pull on a rubber band the strings get straighter. But these strings are being bombarded on the side by these other atoms trying to shorten them by kicking them. So it pulls back—it's trying to pull back, and this pulling back is only because of the heat.

Richard Feynman

8.1 SINGLE-MOLECULE SPRINGS: ORIGINS OF MOLECULAR ELASTICITY

Chapter 2 outlined several methods through which pulling forces can be exerted on individual molecules. This chapter examines, in more detail, what can be learned from such measurements. A parallel between molecules and mechanical springs has already been invoked in Chapter 4 to discuss the behavior of a diatomic molecule: When the length of the bond between its two atoms is changed by a small amount x relative to the equilibrium value, the potential energy is increased by

$$V(x) \approx (1/2)\gamma_0 x^2, \tag{8.1}$$

where the parameter γ_0 is an intrinsic property of the molecule. This potential energy is identical to that of a linear spring with the spring constant γ_0. If we could devise a way to stretch a single diatomic, it would thus generate a restoring force obeying Hooke's law:

$$f = -V'(x) = -\gamma_0 x. \tag{8.2}$$

Eq. 8.2 can be derived by noting that, when the spring extension x is increased by an infinitesimal amount dx, the exerted force $-f$ (which has the direction opposite that of the restoring force) performs a work equal to $dW = -f dx$, which causes the molecule's potential energy to increase from $V(x)$ to $V(x + dx) \approx V(x) + V'(x)dx$. Conservation of energy then requires that

$$-f dx = V'(x)dx,$$

which readily gives Eq. 8.2. The same reasoning, of course, also applies when x is not small and the potential $V(x)$ is no longer quadratic. Therefore if we know $V(x)$ we could always compute the restoring force as its derivative and, conversely, if we knew the dependence of the restoring force on the molecule's extension x we could integrate this dependence to estimate $V(x)$. Pursuing this line of investigation into diatomic molecules is, however, neither realistic nor particularly compelling, as the potentials of diatomics are already known quite accurately from other types of measurements and quantum mechanical calculations. Single-molecule pulling experiments are more commonly employed to study interactions within (or between) larger, polyatomic

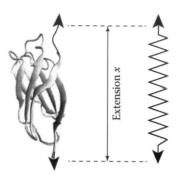

FIGURE 8.1 In this chapter, we will explore the properties of molecules as "springs." By analogy with regular mechanical springs, we will attempt to characterize the relationship between
the molecular extension x and the corresponding restoring force produced by the molecule.

molecules such as DNA, RNA, and proteins. A common property of those is that
they all are chain molecules (i.e., polymers). Their spring action can, for example, be
probed by applying opposing forces at the ends of the molecular chain (Fig.8.1). When
applied to such systems, the above relationship between the potential energy and the
extension is no longer correct. What is missing is a different source of molecular
elasticity that is due to thermal motion.

A simple example of a force induced by thermal motion is the pressure of a gas. A
gas of noninteracting molecules confined to a container of volume v at a temperature
T exerts on the container walls a pressure (i.e., the force per unit area) given by the
ideal gas equation:

$$P = Nk_B T/v. \tag{8.3}$$

Here N is the number of gas molecules and k_B, as before, is Boltzmann's constant.
In contrast to Eq. 8.2, the pressure is proportional to the gas temperature and would
vanish in the (hypothetic) limit of ceasing thermal motion, $T \to 0$. We will return to
the ideal gas example later in this chapter to seek further insight into the connection
between thermal motion and molecular elasticity. But before doing so, it will be
helpful to derive formal statistical-mechanical relationships that will replace Eq. 8.2
in situations where thermal motion is important.

The atoms belonging to any molecule are subject to incessant random motion
caused by their interactions with the surrounding molecules. As a result, the molecule's
total energy E (equal to its potential energy V plus the kinetic energy) is no longer
fixed. Rather, it is a fluctuating quantity whose probability is proportional to $\exp(-\beta E)$
(see Appendix B), where, for convenience, we have defined

$$\beta = (k_B T)^{-1}.$$

Let us enumerate the possible energy values using an auxiliary index i:

$$E = E_i, i = 1, 2, \ldots.$$

In a quantum mechanical description, molecular energies may take on discrete values
while in classical mechanics any energy value is possible. In the latter case the index

i should be viewed as a continuous, rather than discrete, variable. While the outcome of our derivation does not depend on whether i is discrete or continuous, the former case will be adopted for notional convenience.

Imagine an experiment where the distance x between a specified pair of atoms of the molecule is increased. We would like to find the average restoring force f that ensues. The possible values of the molecule's energy depend on the distance x,

$$E_i = E_i(x), i = 1, 2, \ldots.$$

If the molecule is found to be in some particular state i, the restoring force is given by an expression analogous to Eq. 8.2:

$$f_i(x) = -dE_i(x)/dx.$$

The probability that the molecule is in state i is given by the Boltzmann weight

$$w_i(x) = e^{-\beta E_i(x)} \bigg/ \sum_i e^{-\beta E_i(x)} \tag{8.4}$$

and so the mean restoring force can be estimated as the weighted average,

$$f(x) = - \sum_i w_i [dE_i(x)/dx] = -dG(x)/dx, \tag{8.5}$$

where the quantity $G(x)$ is defined by

$$e^{-\beta G(x)} = \sum_i e^{-\beta E_i(x)}. \tag{8.6}$$

We call $G(x)$ the molecule's free energy (viewed as a function of the distance x) or the potential of mean force.

The function $G(x)$ has a very useful physical interpretation. Consider the following ratio:

$$w(x) = \frac{\sum_i e^{-\beta E_i(x)}}{\sum_{i,x} e^{-\beta E_i(x)}}.$$

Since $e^{-\beta E_i(x)}$ is proportional to the probability that the molecule has the extension x and is in the state i, it is clear that $w(x)$ is the probability that the extension is equal to x.[1] Comparing this result to Eq. 8.6, we conclude that

$$G(x) = -k_B T \ln w(x) + \text{constant.} \tag{8.7}$$

The value of the constant is insignificant since it disappears upon differentiation in Eq. 8.5. The result we have obtained is very powerful: Through Eq. 8.5, it relates the zero-force, equilibrium probability distribution of the extension x to the molecule's mechanical response. Note that a one-dimensional particle subjected to a potential

---

[1] For notational symmetry, x is treated as a discrete variable here. In the continuous case, the sum over x should be replaced by an integral. The function $w(x)$ then becomes the probability density of x.

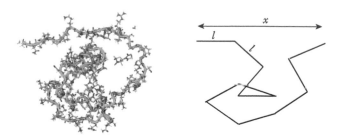

FIGURE 8.2 Long chain molecules (such as the intrinsically disordered protein shown on the left) often adopt random, spaghetti-like shapes. The freely jointed chain (FJC) model shown on the right is often invoked to model such molecules. FJC conformations are random walks in space, whose mathematical properties are described in Appendix A.

$V(x)$ and obeying the canonical distribution at temperature T would have a probability density $w(x)$ given by $w(x) \propto \exp[-V(x)/k_B T]$; comparison of this with Eq. 8.7 immediately gives $V(x) = G(x)$ (to within an insignificant additive constant). Thus free energy becomes identical to the potential energy in the one-dimensional case.

To sum up our findings so far, the restoring force generated by a molecule is governed by its free energy $G(x)$ rather by its potential energy (with the two becoming identical in 1D). That is, as far as their mechanical properties are concerned, molecules can be treated as mechanical springs (cf. Eq. 8.2) provided that the effective potential $G(x)$ is used instead of the actual $V(x)$ in each case.[2]

To illustrate the difference between the potential energy $V(x)$ and free energy $G(x)$, consider the mechanical behavior of a long, completely disordered chain molecule (Fig. 8.2). Although pulling at the ends of a piece of a long macroscopic string (dis)arranged in this fashion would produce no restoring force (until the string becomes taut), thermally agitated microscopic strings exhibit elasticity even when they store no elastic energy. Indeed, according to Eq. 8.5, zero restoring force would imply that the free energy $G(x)$ and, consequently, the probability density $w(x)$ are independent of x, which cannot be true in view of the following argument. Conformations of a flexible molecule resemble trajectories traced by a random walker exploring three-dimensional space (see Section 4.4 and Appendix A). A large value of x would then correspond to a trajectory where the walker happens to step in the same direction most of the time, which is highly improbable given the random direction of each step. We can make this random walk model of a polymer quantitative by assuming

---

[2] One other subtle difference from mechanical springs should be pointed out here: Generally speaking, both the force and the extension should be vectors in the three-dimensional space. If the two are perfectly aligned with each other, they can be replaced by scalar quantities, which is why we do not usually use vectors in writing Eq. 8.2. But in the molecular world, thermal motion will always prevent perfect alignment between the two vectors and so the use of the absolute values of f and x in Eq. 8.5 would not be correct. Instead, both quantities should be treated as vectors, unless the force f is high enough to suppress thermal fluctuations in the transverse direction with respect to the force. In the following, x is always taken to be the component of the molecular extension vector measured along the direction of the force and so Eq. 8.5 remains valid, regardless of whether or not the molecule can be considered well aligned with the force.

some specific length l for the walker's step. The polymer is then effectively viewed as a "freely jointed chain" (FJC) of stiff rods of length l (see Fig.8.2), which are free to rotate around the hinges joining them.[3] Using results from Appendix A, the probability distribution of the x-component of the vector that connects the start and end points of the walk can be estimated as a Gaussian:

$$w(x) = \left(\frac{3}{2\pi nl^2}\right)^{1/2} \exp\left(-\frac{3x^2}{2nl^2}\right),$$

where n is the chain length (i.e., the number of rods forming the chain). Substitution of this into Eq. 8.7 gives

$$G(x) = \frac{3k_B T x^2}{2nl^2} + \text{constant},$$

where the value of the constant is independent of the chain extension. Using Eq. 8.5, we find our polymer to obey Hooke's law,

$$f = -\frac{3k_B T x}{nl^2}, \tag{8.8}$$

with a spring constant, $\gamma = 3k_B T/(nl^2)$, that is proportional to the temperature T.[4] This property is a signature of the molecular elasticity caused by thermal motion. Indeed, should the thermal motion cease (as in the limit $T \to 0$), the spring constant would vanish.

It does not take sophisticated experiments with tiny molecular springs to appreciate the consequences of this elasticity mechanism. It, for example, bestows one of the most common materials, rubber, with a rather unusual property—the propensity to contract upon heating. The microscopic structure of rubber involves numerous random polymer chains chemically cross-linked to one another. As a result, rubber behaves (mechanically) as a random network of springs, each with a spring constant proportional to temperature. If a weight is suspended on a rubber string, heating the rubber (and thus increasing the material's spring constant) will cause the string to contract thereby lifting the weight (and performing the work against gravity).

The result expressed by Eq. 8.5 implies that the work $-f dx$ performed by the exerted force (equal to the restoring force f taken with the opposite sign) is equal to the change of the molecule's free energy $dG = G(x+dx) - G(x)$. Equivalently, in a process where the free energy decreases, $G(x + dx) < G(x)$, the restoring force can

---

[3] Because two atoms of a polymer chain cannot be present at the same spatial location, a self-avoiding random walk (i.e., one that cannot cross itself) would be a better model. While mathematically more complicated, the latter model yields predictions similar to those of the FJC.

[4] As noted in Appendix A, the Gaussian approximation for the probability distribution of the extension x breaks down when x is sufficiently large. Accordingly, Hooke's law expressed by Eq. 8.8 becomes inapplicable at large forces, as the polymer extension approaches its maximum possible value of nl and the chain becomes inextensible. A more accurate treatment of the FJC model described at the end of this chapter (Section 8.5) gives a nonlinear relationship between f and x, which reduces to Eq. 8.8 at low forces/extensions. It should be noted that most single-molecule studies of proteins and DNA are performed in the "high" force regime where nonlinearity of the function $f(x)$ cannot be neglected.

FIGURE 8.3 When the piston moves, the energy of the ideal gas kept at a constant temperature T does not change. Nevertheless, the force f exerted by the gas on the piston can perform mechanical work to lift it. The energy required to accomplish this process thus originates not from the internal energy of the gas but from the heat absorbed from the surroundings.

perform a positive work equal to $f dx$. Thus free energy stored in a molecule can be utilized to perform mechanical tasks in much the same as the energy of a wound up spring can power an alarm clock. These observations appear to sound like statements of energy conservation except that they are not because the quantity G *is not* energy.

Let us take a closer look at the energy balance in a process where the molecule's extension is reduced by dx, resulting in a lower value of the free energy. A positive work $dW = f dx$ is performed by the restoring force as a result. When, for example, all the little microsprings within a rubber string simultaneously contract, the ensuing work can be used to lift a weight. But where does the energy necessary for this come from? Assuming the FJC description for each individual polymer within rubber, the total energy E of the material does not change, i.e., $dE = 0$. The weight then appears to be lifted at no *energy* cost. At the same time, it appears to have costed *free energy*, yet we must remember that fundamental laws of physics require conservation of energy and not free energy. The only possible answer is that the required energy is supplied by the surrounding molecules. If we call the energy exchanged with the surroundings dQ then energy conservation can be saved provided that the following relationship is satisfied:

$$f dx = -dE + dQ \qquad (8.9)$$

(where dE vanishes in the above example but is not necessarily zero for other materials). Using 8.5, we also find:

$$dG = dE - dQ. \qquad (8.10)$$

As another example of utilizing the energy of the surroundings to perform mechanical work, consider the setup shown in Fig. 8.3, where an ideal gas expands, at constant temperature, lifting a piston and performing a positive work dW. Just like for our flexible chain molecules, we have $dE = 0$ because the energy E of an ideal gas only depends on its temperature T. The energy required to lift the weight, dW, must then entirely originate from the heat dQ absorbed by the molecules of the gas from its surroundings and required to keep the temperature of the gas constant.

Although the quantity dQ was introduced above to salvage the energy balance, its existence also follows from the formal mathematical arguments used to derive Eq. 8.5. Specifically, a change in the extension x causes all energies $E_i(x)$ to change.

Consequently the Boltzmann probability, Eq. 8.4, of finding the molecule in each state changes as well. This rebalancing of the populations w_i can only happen as a result of transitions among different energy states E_i; conservation energy requires that every time such a transition, say to a higher energy state, takes place, the energy difference must be supplied by the surrounding molecules. Therefore, maintaining proper thermal distribution among various energy states of the molecule necessitates energy flow between the molecule and its surroundings.[5]

The average amount of energy, or heat dQ, received by the molecule from its surroundings can be further calculated using Eqs. 8.10 and 8.6. Noting the identity $E = \sum_i w_i(x)E_i(x)$ for the mean energy of the molecule, we write

$$dQ = dE - dG = \sum_i (dw_i E_i + w_i dE_i) - (1/\beta)d \ln \sum_i e^{-\beta E_i} = \sum_i E_i dw_i. \tag{8.11}$$

Introducing the new quantity

$$S = -k_B \sum_i w_i \ln w_i = k_B \left\langle \ln \frac{1}{w_i} \right\rangle, \tag{8.12}$$

it is now easy to see that

$$dQ = TdS. \tag{8.13}$$

Indeed, this result is immediately obtained by differentiating Eq. 8.12 and noting that all the terms proportional to $d \sum_i w_i = \sum_i dw_i$ are zero because the probabilities w_i always add up to one.

The quantity S is called entropy. To get a better feel for it, imagine that all the probabilities w_i are the same. Then Eq. 8.12 becomes

$$S = k_B \ln \Omega, \tag{8.14}$$

where $\Omega = 1/w_i$ is the number of states i. The FJC model (Fig.8.2) exemplifies this scenario, with the number of states $\Omega(x)$ simply equal to the number of different (and equally probable) random walks consistent with the specified value of x.[6] Moreover, the probability $w(x)$ in this case is equal to $\Omega(x)$ normalized by the total number of walks. When x is increased, the chain is driven to a "less likely" state with a smaller value of the probability $w(x)$ and, therefore, with a lower entropy S. Using Eqs. 8.11

---

[5] The picture will, however, change if the experiment is carried out so fast that thermal equilibrium within the molecule cannot keep up with the changes in x. Such nonequilibrium processes will be discussed in the next chapter.

[6] Strictly speaking, Ω is infinite for any classical system, whose state is described by continuous variables (coordinates, angles, etc.). To avoid this difficulty, we could, e.g., consider a discretized FJC version, in which the angle between two adjacent rods can only take on a discrete set of values. The number of states Ω and, consequently, the entropy defined by Eq. 8.14 clearly depends on the number of allowed values of this angle. However, the change of the entropy, $\Delta S = k_B \ln \Omega(x_2)/\Omega(x_1)$, in a process where the distance x, say, increases from x_1 to x_2 is independent on how the FJC model is discretized because the use of a finer resolution model amounts to multiplying the number of states Ω by a constant number. In most physical problems, the absolute value of entropy is irrelevant and only its changes matter.

and 8.13, we can now write $dG = dE - dQ = dE - TdS = d(E - TS)$, which suggests that the free energy can be written in the form

$$G(x) = E(x) - TS(x). \tag{8.15}$$

Consequently, the restoring force,

$$f = -dG/dx = -dE/dx + TdS/dx, \tag{8.16}$$

has two contributions, one from the molecule's energy and the other from its entropy. This result implies that mechanical work is required not only in order to increase the molecule's energy but also to reduce its entropy, i.e., to drive it towards a state that is less likely to be encountered in equilibrium in the absence of the force. The quantity

$$TdS/dx$$

is often called "entropic force." The restoring force exerted by a random polymer, Eq. 8.8, is of an entirely entropic origin, as the energy of the chain does not change. The pressure, Eq. 8.3, exerted by an ideal gas on a piston in Fig.8.3 is another example of a purely entropic force because the energy E of an ideal gas depends only on its temperature T and so the work exerted, say, to compress the gas at a constant temperature must be entirely expended to reduce its entropy. Despite its wide acceptance, the term "entropic force" is somewhat misleading for it appears to suggest that entropy is some metaphysical force-causing entity. From a microscopic perspective, however, the pressure of the gas is produced by the collisions of the individual gas molecules with the walls and so it, ultimately, is a cumulative effect of simple, mechanical forces. Likewise, a chemist or a physicist using a computer to perform molecular simulation of a random polymer chain could estimate the average restoring force f and its dependence on its extension x directly, using Newtonian laws of mechanics applied to the atoms of the polymer and its surrounding molecules, without ever knowing anything about entropy or free energy. The power of the thermodynamic quantities such as free energy and entropy, however, is that they allow, through Eq. 8.5, to derive equations like Eq. 8.3 or 8.8 and to predict the cumulative effect of a large collection of atoms and/or molecules without having to solve for the trajectory of each.

8.2 THERMODYNAMICS AND KINETICS OF MECHANICALLY RUPTURED BONDS

Equation 8.5, as derived, determines the average restoring force produced by a molecule when its extension x is clamped at a given constant value. Having the full control of the microscopic variable x is, however, unfeasible in most experimental studies. Rather, some dynamometer device is required both to transmit the force to the molecule and to measure this force. Most molecule pulling devices are spring-type dynamometers. For example, in the atomic force microscope (AFM), the force exerted on an AFM tip causes deflection of a cantilever (Fig. 2.2), which is proportional to the force. In an optical tweezer setup the spring action is supplied by the radiation force acting on a polystyrene bead that is attached to the molecule and

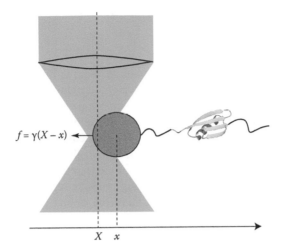

FIGURE 8.4 When the center of the optical trap (X) is shifted relative to the location of the bead x, this creates a force toward the trap center, which is proportional to $X - x$. The coupling of the "molecular" coordinate x and the "instrument" coordinate X can then be modeled by a harmonic potential $\gamma(X - x)^2/2$, where the trap spring constant γ describes how soft (or stiff) the trap is.

placed within an optical trap (Fig.8.4). As the experimenter changes the trap position X at will, the trap pulls on the molecule with a force $\gamma(X - x)$, where γ is the trap stiffness. A key difference from force measurements performed on the macroscale is then that in single-molecule pulling experiments both the molecular extension x and the trap force are subject to appreciable thermal fluctuations, whose magnitudes depend on the properties of both the molecule and the pulling device.

We will model the pulling device by introducing a harmonic potential $\gamma(X - x)^2/2$, where X is the experimentally controlled instrument variable (such as the trap position) and x is the molecular extension.[7] The total free energy of the molecule coupled to the trap is then a function of both X and x

$$\tilde{G}(X, x) = G(x) + \gamma(X - x)^2/2. \qquad (8.17)$$

In the limit of a very stiff trap (large γ) the high energetic penalty associated with the trap necessitates that we have

$$x \approx X$$

and

$$\tilde{G}(X, x) \approx G(X).$$

Thus the extension is clamped at the desired value X, as assumed in the previous section. This scenario, however, is rarely attained in practice. The opposite extreme

---

[7] Small molecules are often attached to the pulling device via relatively long molecular handles (often DNA). If we wish to study the properties of such a molecule alone (i.e., without the handles) then γ must be defined as the combined spring constant of the optical trap and the handles. Note, however, that it is often necessary to go beyond Hooke's law to describe elasticity of molecules such as DNA.

of a very soft pulling device is, in contrast, often closer to reality. Let us rewrite Eq. 8.17 as

$$\tilde{G}(X, x) = G(x) + \gamma X^2/2 - \gamma Xx + \gamma x^2/2 \tag{8.18}$$

and suppose, for a moment, that the free energy $G(x)$ is that of a harmonic spring (Eq. 8.1), i.e., $G(x) - \gamma_0 x^2/2$, where the spring constant γ_0 describes the intrinsic molecular stiffness. The soft pulling device assumption then amounts to the inequality $\gamma_0 \gg \gamma$. Suppose now that a constant value of the control variable X is maintained and consider the free energy experienced by the molecule as a function of x. The last term in Eq. 8.17 can be safely disregarded as it is negligible when compared to the much larger quadratic term $G(x)$. The effect of the trap is now contained in the term that is linear in X and x. This term is nonnegligible provided that X is sufficiently large, particularly much larger than x. We, therefore, approximate our free energy by

$$\tilde{G}(X, x) \approx G(x) - fx + \gamma X^2/2,$$

where

$$f = \gamma X$$

is the force exerted on the molecule by the trap. The molecule sees the x-dependent part of this potential,

$$G_f(x) = G(x) - fx. \tag{8.19}$$

This force-modified free energy contains the additional term $-fx$, which accounts for the molecule's interaction with a constant stretching force f and is analogous to the potential energy mgh of a weight of mass m subjected to the gravitational force $f = -mg$ when placed a height h above the ground.

The corresponding probability distribution of the extension x is given by

$$w_f(x) = Ae^{-\beta G_f(x)},$$

where A is a normalization factor. The maximum of this distribution is achieved at a point x_m satisfying the equation $dG_f/dx = G'(x_m) - f = 0$ or

$$f = G'(x_m), \tag{8.20}$$

which is the same as Eq. 8.5 aside from the sign convention (i.e., here f is the force applied to the molecule). These two equations, however, describe different experimental scenarios. Equation 8.5 corresponds to the hypothetical scenario where the extension x is fixed and the *average* of a fluctuating force is measured. Equation 8.20 describes the case where a constant (i.e., non-fluctuating) force is imposed by the instrument. It then predicts the *most probable* value of the fluctuating extension x. But this is not what one would normally expect to measure: Rather, the *mean* position,

$$\langle x \rangle = \int_{-\infty}^{\infty} dx x w_f(x),$$

would be the expected outcome. In general, there is no reason for the mean value of a quantity to be exactly the same as the point where its probability density attains the maximum. We, therefore, arrive at the surprising conclusion that the two experiments

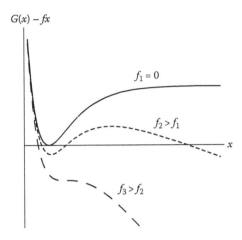

$G(x) - fx$

$f_1 = 0$

$f_2 > f_1$

x

$f_3 > f_2$

FIGURE 8.5 While the zero-force effective potential $G(x)$ shown here exhibits a global minimum, a pulling force f destabilizes the bound state of a complex so that states where its fragments are separated by a large distance are always favored thermodynamically. A low enough force f_2 renders the bound state metastable (i.e., $G_f(x)$ exhibits a local minimum), while a higher force f_3 eliminates the bound state altogether.

will not necessarily yield the same force-extension curve! Such a discrepancy does not arise in measurements performed on macroscopic objects because fluctuations in either the extension or the force are negligible.

Exercise

Derive the dependence of the mean extension $\langle x \rangle$ on the applied force f assuming that $G(x) = \gamma_0 x^2/2$. Compare your result with Eq. 8.5.

Our discussion so far did not allow for the possibility that a molecular spring can break. This possibility, however, naturally arises from considerations of the free energy profiles arising from inter- or intramolecular cohesive interactions or bonds. The simplest example of such a bond is offered by a diatomic molecule, with the dependence of the interaction energy on the interatomic distance depicted in Fig.4.1. We may then envisage a thought experiment where a force is used to pry this molecule apart. A more realistic scenario is offered by forced separation of a protein and its ligand or of two proteins involved in a complex that is responsible for inter-cell adhesion. The physical meaning of x is now the distance between the two fragments within the complex. The function $G(x)$ displays a minimum and the probability density $w(x) \propto \exp[-G(x)/k_BT]$ displays a maximum corresponding to the most probable value of the inter-fragment separation. When x is increased, $G(x)$ rises monotonically, leveling off at a finite value that represents the free energy cost of breaking the adhesive bond between the fragments (see Fig. 8.5, the case of zero force). For any finite value of the force, however, we have $G_f(x) \to -\infty$ as $x \to \infty$. This means that the probability $w_f(x) \propto \exp[-G_f(x)/k_BT]$ grows indefinitely as

$x \to \infty$ and so the ruptured complex with its fragments separated is always favorable, thermodynamically, as compared to the complex that is intact. Two distinct physical regimes are possible in this case. For a sufficiently low force (see, e.g., the case $f = f_2$ in Figure 8.5) there is a barrier separating the local minimum and the maximum of the effective potential $G_f(x)$. The corresponding probability $w_f(x)$ still has a *local* maximum corresponding to the minimum of $G_f(x)$. However upon further increase in x, this probability will eventually increase with the separation. Starting near its equilibrium state (the minimum of $G_f(x)$), the complex will appear stable with respect to fluctuations in the distance x that are sufficiently small. That is, small decreases or increases in x will lead to less probable configurations (i.e., smaller values of $w_f(x)$). A sufficiently large increase in x, however, will take the molecule accross the barrier in $G_f(x)$ and lead to dissociation of the complex toward larger values of x. States corresponding to local minima of $G(x)$ are known as "metastable": they may appear to be stable at sufficiently short times (corresponding to small fluctuations) but are eventually abandoned in favor of the more likely configurations (i.e., the dissociated complex in the present case).

As the force is increased to a sufficiently high value (e.g., $f = f_3$ in Fig.8.5), the second regime is encountered, where the local minimum and the barrier separating it from the lower free energy states disappear altogether. This means that the probability $w(x)$ grows monotonically as the separation x is increased and the complex is unstable, even with respect to small fluctuations. Applying such a high force will, therefore, lead to immediate dissociation of the complex.

The first of these two regimes is more interesting and involves longer, experimentally accessible timescales. It belongs to the realm of chemical kinetics and can be viewed as an example of an irreversible chemical reaction

$$\text{METASTABLE COMPLEX} \xrightarrow{k(f)} \text{FRAGMENTS.}$$

The rate coefficient $k(f)$ can be measured directly as the inverse of the mean time it takes the complex to rupture at a given force f (see Chapter 3). The underlying free energy $G(x)$, on the other hand, is often inaccessible to a direct measurement, despite its simple connection to the equilibrium distribution $w(x)$. The problem is that $w(x)$ often varies substantially upon very small, subnanometer-scale changes in the distance x, a spatial resolution not afforded by most experimental setups. It is then customary to deduce the sought after function $G(x)$ from the experimental dependence $k(f)$. How can this be done?

A simple (yet—as will be found later—sometimes inadequate) solution is suggested by the analogy between $G(x)$ and the potential energy $V(x)$, the two quantities that would be identical had the molecule of interest had only one degree of freedom x. Treating $G(x)$ as an effective potential, one can follow the reasoning from Chapter 4 to construct a model for the dynamics of the extension x. This, for instance, could be a Langevin equation describing a one-dimensional particle in a potential $G(x)$ (or $G_f(x)$ in the presence of a force f). Such a Langevin equation would automatically recover the correct *equilibrium* distribution $w_f(x) \propto \exp[-G_f(x)/k_B T]$ even though there is, of course, no guarantee that it will correctly capture the rupture *dynamics*. If such a leap of faith is accepted, the approaches developed in Chapter 5 can then be utilized to estimate the reaction rate $k(f)$. Regardless of whether one uses transition

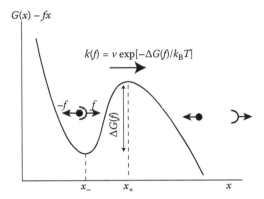

FIGURE 8.6 Rupture dynamics of a molecular adhesion bond can be described as the escape from a metastable state, with a rate coefficient $k(f)$ whose force dependence is primarily controlled by the height of the barrier $\Delta G(f)$. The barrier can be computed as a difference between the values of $G_f(x)$ at its maximum x_+ and the local minimum x_-.

state theory (TST) or other approaches (such as Kramers' theory) that improve TST estimates, the final answer can be written in the form:

$$k(f) = v e^{-\frac{\Delta G(f)}{k_B T}}, \tag{8.21}$$

where $\Delta G(f)$ is the free energy difference between the top of the barrier and the local minimum of $G_f(x)$ (see Fig. 8.6). Although the prefactor v is, strictly speaking, itself dependent on the force, this dependence tends to be weak as compared to the *exponential* dependence caused by the force induced variation of the free energy barrier $\Delta G(f)$. We will thus treat v as a constant.

Let $x_+ = x_+(f)$ be the position of the maximum and $x_- = x_-(f)$ the position of the minimum of $G_f(x)$. Then we have

$$\Delta G(f) = G(x_+) - G(x_-) - f(x_+ - x_-).$$

Differentiating this with respect to the force, we find

$$d\Delta G/df = G'(x_+)(dx_+/df) - G'(x_-)(dx_-/df) - (x_+ - x_-)$$
$$- f[(dx_+/df) - (dx_-/df)].$$

This expression becomes greatly simplified after we notice that the extrema of the function $G_f(x)$ satisfy the condition $dG_f(x)/dx = 0$ (cf. Eq. 8.20) or, equivalently,

$$G'(x_\pm) - f = 0,$$

leading to the cancellation of all the terms proportional to $dx_\pm/df$. Therefore,

$$d\Delta G/df = -(x_+ - x_-) \tag{8.22}$$

and

$$\Delta G(f) = \Delta G(0) - \int_0^f [x_+(f') - x_-(f')]df'. \tag{8.23}$$

At sufficiently low values of the force f, the positions $x_\pm(f)$ can be approximated in Eq. 8.23 by their zero-force values, which gives

$$x_+(f) - x_-(f) \approx x_+(0) - x_-(0) \equiv \Delta x$$

and

$$\Delta G(f) \approx \Delta G(0) - f\Delta x.$$

Substituting this linear approximation into Eq. 8.21, we arrive at the simple formula

$$k(f) = k(0)e^{\frac{f\Delta x}{k_B T}}, \tag{8.24}$$

which predicts exponential growth of the rate with the force. This result, popularized in the famous article by Bell [1], is commonly known as the Bell formula, although it had been introduced earlier by Zhurkov in the context of fracture mechanics [2] and, in fact, was mentioned by one of the pioneers of transition state theory, Henry Eyring, in the 1930's [3]. When consistent with measured $k(f)$, Eq. 8.24 allows experimenters to interpret their measurements in terms of two numbers, the zero-force rate coefficient $k(0)$ and the location Δx of the barrier relative to the initial state. Although not as detailed as the function $G(x)$, Eq. 8.24 provides a simple and useful view of force-induced bond rupture.

Exercise

Derive the expressions describing the force dependence of $k(f)$ for the following two potentials:

$$G(x) = \gamma_0 x^2 \left(1 - \frac{x}{x_0}\right) \Big/ 2$$

and

$$G(x) = \begin{cases} \gamma_0 x^2/2, \, x \le x_0 \\ -\infty, \, x > x_0 \end{cases}.$$

Compare the results with the Eyring-Zhurkov-Bell formula.

8.3   SLIP VS. CATCH BONDS

Despite its simplicity, Eq. 8.24 is remarkably successful and is routinely used to interpret experimental data. Some molecular bonds, however, exhibit very different behavior, with their lifetime *increasing* upon application of a force. Those are called "catch bonds," as opposed to the more common "slip bonds" that were discussed in the preceding section. The catch-bond behavior could, of course, be formally reproduced with Eq. 8.24 if the distance Δx were allowed to be negative, but the physical origin of a negative Δx is not immediately clear. Indeed, a $k(f)$ that decreases with increasing force would imply a free energy barrier that grows with the force, which would be at odds with the model depicted by Fig. 8.5.

Before discussing what might be wrong with that model and how it could be fixed, let us note that catch-bond-like behavior is common in the macroscopic world. Use

FIGURE 8.7 Two interlocked hooks provide a macroscopic prototype of a catch bond.

of door wedges or sailing knots, for example, relies on jamming or tightening that increases their resistance to the applied force. An even simpler yet suggestive example is provided by two interlocked hooks shown in Figure 8.7. If pulled away along a straight line, the hooks cannot be separated until they break or bend. The successful unhooking motion cannot be executed by moving along a straight line or induced by application of a constant force. The description of such motion clearly requires more than one degree of freedom—Hooks do not even exist in one dimension! This suggests that the low dimensionality of our earlier model (Fig. 8.6) may be the reason why it fails to reproduce the counterproductive effect of the pulling force.

Indeed, it is not hard to construct a two-dimensional model (Figure 8.8) that will exhibit a catch-bond behavior. In this model, the free energy depends on two variables, the mechanical extension x and some other internal variable y representing the remaining molecular degrees of freedom, and can be defined in a manner similar to Eq. 8.7 as:

$$G(x, y) = -k_B T \ln w(x, y) + \text{constant},$$

where $w(x, y)$ is the joint probability distribution of the two variables. The force-modified free energy landscape is given by $G_f(x, y) = G(x, y) - fx$, where the second variable does not couple to the mechanical force. As in the one-dimensional case, $G_f(x, y)$ exhibits a minimum corresponding to the adhesion bond in its intact form (provided that the force f is not too high). The rate of escape from this minimum requires crossing a "mountain ridge" and is dominated by the trajectories that pass in the vicinity of a saddle point (shown as the cross in Fig. 8.8), which provides the lowest (free) energy escape route (see Chapter 5). The catch-bond behavior now

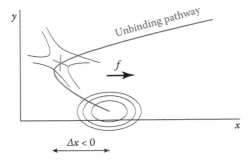

FIGURE 8.8 A minimal model showing the catch-bond behavior consists of a two-dimensional free energy landscape, as a function of the pulling coordinate x and another collective degree of freedom, y, representing the remaining degrees of freedom. Pulling along x diverts the system away from the energetically preferred unbinding pathway (red line) thereby making the lifetime of the adhesion bond longer.

arises as a consequence of the circuitous nature of this dominant pathway, which is reminiscent of the unhooking motion: It requires the extension to first decrease ($\Delta x < 0$) before it eventually increases when the components of the complex become separated. A stretching force ($f > 0$), while favoring unbinding thermodynamically, has a counterproductive *dynamic* effect as it initially pushes the system away from its favorable pathway thus increasing barrier and the corresponding bond lifetime.

It should be emphasized that nothing is technically wrong with the one-dimensional free energy $G_f(x)$ (Fig. 8.5) *per se*, but the free energy is an *equilibrium* property, which is not guaranteed to tell us anything about time-dependent phenomena. On the other hand, the assumption that the *dynamics* along x can be described as simple motion in the one-dimensional potential $G(x)$ (such as Langevin dynamics or transition state theory) has no sound foundation and is clearly inadequate when it comes to catch bonds. Another way to think about the failure of the one-dimensional model is to consider Eq. 8.21. According to the results from Chapter 5, this equation is always technically true provided that the prefactor ν includes the appropriate transmission coefficient. It was tacitly assumed that this prefactor is sufficiently well behaved and, in the very least, is insensitive to the force f so that the effect of the force is entirely contained in the force dependence of the one-dimensional free energy barrier. Although this assumption is justifiable for simple models implied by Kramers' theory or transition state theory, we had no real grounds to generalize it to all force-dependent rupture phenomena.

By the same token, adequacy of the two-dimensional model cannot be guaranteed. It, nevertheless, appears physically reasonable in that it provides one with enough flexibility to describe both catch and slip bonds. A further argument in favor of 2D as the lowest physically acceptable dimensionality of a model to describe the coupling of chemical (e.g., bond breaking) and mechanical phenomena is that one degree of freedom is required to describe each. That is, the choice of x as the extension is dictated by the necessity to describe the coupling of the molecule to the mechanical force; except, perhaps, by accident, there is no reason to expect that x would be also a good variable to describe the inherent dynamics of bond breaking. As a result, at least one more degree of freedom, y, is needed.

8.4 FORCE-INDUCED UNFOLDING AND OTHER CONFORMATIONAL TRANSITIONS INFLUENCED BY FORCES

The discussion so far was limited to the irreversible unbinding process. Mechanical forces can also be used to control reversible processes such as biomolecular folding and unfolding, as illustrated in Fig.8.9. Cohesive interactions now occur between parts of the same molecular chain rather than between two different entities. Similarly to the case of an adhesive bond, those interactions lead to a minimum in the free energy $G(x)$ viewed as a function of the distance x between the parts of the chain (typically, its ends) at which the forces are applied. Unlike the case of irreversible unbinding, however, the free energy $G(x)$ no longer reaches a constant value at $x \rightarrow$

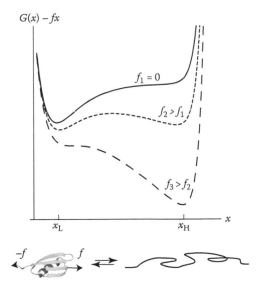

FIGURE 8.9 A mechanical force applied to the ends of the polypeptide chain can be used to shift the thermodynamic equilibrium between folded and unfolded states of a protein. At finite values of the stretching force f, the free energy profile $G_f(x) = G(x) - fx$ may show two distinct minima, one corresponding to the folded state and the other to the unfolded state. Because of the term $-fx$, the state that is less compact (i.e., corresponding to a greater value of x) is favored at higher forces, resulting in an unfolded state minimum whose depth increases with the force f. If the folded state is thermodynamically favorable at zero force, it is possible to tune the value of the force f such that the folded and the unfolded states are equally populated. Observation of real-time dynamics under such a force will reveal that the protein extension x will jump between high (H) and low (L) values (cf. Chapter 4).

∞ but, instead, rises sharply when the molecular chain becomes highly extended.[8] A high value of the pulling force will obviously unfold and stretch the chain. Thus thermodynamic equilibrium may be shifted from the folded to the unfolded state by increasing the force. An intermediate value of the force should further exist, where the unfolded and folded states are equally populated. When the molecule is held at such a force, its extension x will exhibit reversible jumps between low and high values, thus manifesting reversible folding/unfolding dynamics. For an example of experimental realization of such a scenario, see ref. [4].

Of course, depending on the underlying intramolecular interactions as well as on the value of the force, more than two minima (and, consequently, more than two distinct states) may be observed in the function $G_f(x)$, resulting in multiple observable states. Similarly to fluorescence measurements described in Chapter 7, identification of such states presents a considerable challenge because the actual signal $x(t)$ tends to be noisy, owing both to the unavoidable thermal fluctuations and to experimental errors. Approaches to data analysis discussed in Chapter 7 are, likewise, applicable

---

[8] The possibility of the chain rupture is not considered here.

to single-molecule mechanical experiments.

The ideas developed in this chapter can be extended to describe other scenarios involving mechanical forces. One example is presented by nanopore experiments discussed in Chapter 2 (see Figure 2.6), where the mechanical force is created by an electrical field acting on the charged atoms of the molecule that, as a result, is driven across a narrow pore. Assuming, for simplicity, a constant electric field E acting along the z-axis, the interaction energy of the field with the molecule takes the form

$$-E \sum_i q_i z_i,$$

where q_i and z_i are, respectively, the charge and the z coordinate of the i-th atom. This term can be written in the same form as above,

$$-fx,$$

where

$$f = E \sum_i q_i$$

is the total electric force on the molecule and

$$x = \sum_i q_i z_i \bigg/ \sum_i q_i$$

is the effective coordinate describing the coupling of the molecule to this force.

The simplest description of the dynamics of the molecule's passage through a nanopore would then involve motion in a force-modified potential $G_f(x) = G(x) - fx$, where the free energy $G(x)$ is related to the equilibrium distribution of the coordinate x through Eq. 8.7. Unlike stretching experiments, however, nanopore studies often focus on the time it takes the molecule to traverse the pore (during which the ionic current through the pore is blocked or suppressed, allowing observation of those events). That is, they present an example of measuring *transit times* discussed in Chapter 6.

Interactions of a molecule with a pore may change its conformation quite significantly. Indeed, passage through narrow pores was shown to cause separation of DNA strands and to unfold proteins. The deficiencies of a one-dimensional description involving a single coordinate x are particularly apparent when considering such pore-induced conformational rearrangements. Indeed, there is no reason for the charge-weighted average geometric coordinate x to account for large changes of the molecular structure. Nevertheless, the simplicity of the one-dimensional free energy cartoon is often too appealing to discard it as physically inadequate. Perhaps minimalist models invoking a second degree of freedom to account for the internal degrees of freedom of the molecule will find increased use in the nanopore field in the future.

8.5 FURTHER DISCUSSION: ELASTIC RESPONSE OF A FREELY JOINTED CHAIN BEYOND HOOKE'S LAW

In Section 8.1 we found that a freely jointed chain (FJC) stretched by pulling on its ends obeys Hooke's law, provided that the force-induced extension is much smaller than the length of the fully stretched chain. In single-molecule pulling experiments (particularly those involving the stretching of DNA and unfolded proteins), this condition is rarely satisfied and, as a result, most polymers exhibit nonlinear elasticity. Here we will derive a more general relationship between the stretching force and the resulting extension, which will account for such nonlinear effects. Let $\mathbf{r}_i$ $(i = 0, 1, \ldots, n)$ be the successive positions of the FJC random walker. They obviously satisfy the condition

$$|\mathbf{r}_i - \mathbf{r}_{i-1}| = l, \qquad (8.25)$$

where l is the length of a single step. Suppose a force $\mathbf{f}$ is applied to the chain at its end at $\mathbf{r}_n$ and the opposing force $-\mathbf{f}$ is exerted at $\mathbf{r}_0$. Since the FJC has no internal energy, the total energy of the chain is a result of the work done by the external forces and is simply given by

$$E(\mathbf{f}, \mathbf{r}_0, \ldots, \mathbf{r}_n) = -\mathbf{f}(\mathbf{r}_n - \mathbf{r}_0).$$

The mean chain extension, as a function of the applied force, can be computed as the Boltzmann-weighted average

$$\langle \mathbf{r}_n - \mathbf{r}_0 \rangle = \frac{\sum_{\text{configurations}} (\mathbf{r}_n - \mathbf{r}_0) e^{\beta \mathbf{f}(\mathbf{r}_n - \mathbf{r}_0)}}{\sum_{\text{configurations}} e^{\beta \mathbf{f}(\mathbf{r}_n - \mathbf{r}_0)}}.$$

Here, the "summation" is performed over all possible chain configurations (i.e., all possible random walks). Introducing the chain partition function

$$q = \sum_{\text{configurations}} e^{\beta \mathbf{f}(\mathbf{r}_n - \mathbf{r}_0)},$$

it is easy to see that the above equation can be written as

$$\langle \mathbf{r}_n - \mathbf{r}_0 \rangle = \frac{1}{q\beta} \frac{dq}{d\mathbf{f}} = k_B T \frac{d \ln q}{d\mathbf{f}}, \qquad (8.26)$$

so our task will be straightforward to accomplish once the force dependence of the partition function q is known. Since the configuration of the chain is described by the continuous variables $\mathbf{r}_i$, the summation over such configurations must actually be integration. However, the chain connectivity constraint of Eq. 8.25 prevents one from integrating over each of those variables independently. The problem is greatly simplified by using the $n + 1$ variables $\mathbf{r}_0, \mathbf{l}_1, \ldots, \mathbf{l}_n$ instead of the positions $\mathbf{r}_0, \mathbf{r}_1, \ldots, \mathbf{r}_n$ to represent the FJC configuration. Here the "bond vector" variables $\mathbf{l}_i$ are defined by

$$\mathbf{l}_i = \mathbf{r}_i - \mathbf{r}_{i-1}.$$

Using these variables and noting the obvious identity $\mathbf{r}_n - \mathbf{r}_0 = \mathbf{l}_1 + \mathbf{l}_2 + \cdots + \mathbf{l}_n$, we can write the partition function as a product of independent terms:

$$q = C \int d^3 \mathbf{r}_0 d^3 \mathbf{l}_1 \ldots d^3 \mathbf{l}_n e^{\beta \mathbf{f}(\mathbf{l}_1 + \mathbf{l}_2 + \cdots + \mathbf{l}_n)} = CV \left[\int d^3 \mathbf{l} \exp(\beta \mathbf{f} \mathbf{l}) \right]^n.$$

Here $V = \int d^3\mathbf{r}_0$ is the total volume available to the polymer and C is a factor which is independent of the force and which accounts—if we think of our random walk as a representation of a physical model consisting of "beads" connected by bonds—for the integration over the momenta of the beads. Integration over the bond variable $\mathbf{l}$ can now be performed if we introduce the spherical coordinates ($l = |\mathbf{l}|, \theta, \phi$) such that the x, y, and z components of the bond vector are given by

$$l_x = l \sin\theta \cos\phi,$$
$$l_y = l \sin\theta \sin\phi,$$

and

$$l_z = l \cos\theta.$$

If the z-axis is chosen to be aligned with the force[9] then we also have

$$\mathbf{fl} = fl \cos\theta,$$

where f is the absolute value of the force. Since the bond length l is strictly fixed in the FJC, integration is only to be performed over the angular variables:

$$\int d^3\mathbf{l} \exp(\beta \mathbf{fl}) \rightarrow \int_0^{2\pi} d\phi \int_0^{\pi} d\theta l^2 \sin\theta \exp(\beta fl \cos\theta) \propto \left(\frac{\exp(\beta fl) - \exp(-\beta fl)}{\beta fl} \right),$$

where force independent factors were omitted as their presence does not affect Eq. 8.26. This gives, for the partition function q,

$$q \propto \left(\frac{\exp(\beta fl) - \exp(-\beta fl)}{\beta fl} \right)^n.$$

Since the partition function depends on the absolute value of the force and not on its direction, Eq. 8.26 can be further rewritten as

$$\langle \mathbf{r}_n - \mathbf{r}_0 \rangle = k_B T \frac{d \ln q}{df} \mathbf{e},$$

where $\mathbf{e} = \mathbf{f}/f$ is the unit vector in the direction of the force. Thus the end-to-end extension vector is aligned with the force, with its absolute value given by

$$|\langle \mathbf{r}_n - \mathbf{r}_0 \rangle| = k_B T \frac{d \ln q}{df}.$$

Performing the differentiation we finally find:

$$|\langle \mathbf{r}_n - \mathbf{r}_0 \rangle| = k_B T n \left[\frac{l}{k_B T} \coth\left(\frac{fl}{k_B T} \right) - \frac{1}{f} \right].$$

---

[9] Throughout this chapter, the convention has been for the pulling force to act along the x-axis. For the purpose of evaluating the partition function integral using the standard spherical coordinates, however, it is more convenient to align the z-axis with the direction of the force. The result, of course, does not depend on the choice of the coordinate system.

At a force low enough that $fl \ll k_B T$ we can expand the hyperbolic cotangent in a series, $\coth a \approx 1/a + a/3 + \cdots$, to obtain a linear force-vs.-extension relationship (i.e. Hooke's law),

$$|\langle \mathbf{r}_n - \mathbf{r}_0 \rangle| = \frac{n l^2 f}{3 k_B T},$$

which is identical to Eq. 8.8 (note that the restoring force produced by the polymer equals the exerted force taken with the opposite sign). However the FJC extension deviates from that prescribed by Hooke's law when fl becomes comparable to the thermal energy $k_B T$. In particular, in the limit $fl \gg k_B T$ this extension approaches its asymptotic value of nl, which describes a fully extended chain. In other words, the force required to stretch the chain increases faster than the Hooke's law prescription, approaching infinity as the polymer extension approaches its maximum value of nl. The high force limit should not, of course, be viewed as a realistic model of real molecular chains, as it disregards the possibility of covalent bond stretching or breaking.

REFERENCES

1. George I. Bell, "Models for specific adhesion of cells to cells", *Science*, vol. 200, 618-627, 1978.
2. S.N. Zhurkov, "Kinetic concept of strength of solids," *Int. J. Fract. Mech.*, vol. 1, p. 311, 1965.
3. H. Eyring, "Viscosity, plasticity, and diffusion as examples of absolute reaction rates", *J. Chem. Phys.*, vol. 4, p. 283, 1936.
4. J. Liphardt, B. Onoa, S.B. Smith, I.J. Tinoco, and C. Bustamante, "Reversible unfolding of single RNA molecules by mechanical force", *Science*, vol. 292, p. 733, 2001.

9 Nonequilibrium Thermodynamics of Single Molecules: The Jarzynski and Crooks Identities

Certainly a bacterium, which through all its life is tossed around by molecular impacts, will sneer at the statement that heat cannot go over into mechanical motion!.

George Gamow, *One Two Three ... Infinity: Facts and Speculations of Science*

9.1 EXTENSION AND CONTRACTION OF MOLECULAR SPRINGS: ENERGY DISSIPATION AND THE SECOND LAW OF THERMODYNAMICS

The work done on an isolated system is equal to the change in its total energy. Molecules in vacuum or in a gas can sometimes be approximated as isolated systems but such a description is guaranteed to be wrong for molecules in solution or in the solid state. Nevertheless, in the preceding chapter we used ideas from statistical mechanics to modify the above statement of energy conservation and to predict that the (average) work $f dx$ done when a molecule is stretched, mechanically, by a force f equals the change of its free energy $dG = G(x + dx) - G(x)$. This enabled us to further establish the connection between the equilibrium distribution of the extension x and the restoring force generated by the stretched molecule.

Here, the energy balance in molecules driven by external forces will be studied more carefully. Upon closer inspection, the above connection between work and free energy will be found to hold only when energy losses, e.g., due to friction, are negligible, which, in turn, requires that pulling is carried out slowly enough that the molecule is found in thermal equilibrium with its surroundings at any stage of the process. Until almost the very end of the 20th century, physicists settled for replacing the exact equality between work and the free energy change by an inequality, which accounts for the energy wasted in the process.[1] A recent discovery has changed this: Certain exact relationships between work and free energy were found to hold true regardless of any details of the molecule's dynamical behavior and no matter how violently the molecule's equilibrium state is disrupted by the force.

---

[1] Of course, this inequality could be replaced by an equality again if the exact amount of wasted energy is known, but this would require precise knowledge of the dynamics obeyed by the molecule.

To illustrate that the outcome of a pulling experiment depends on how fast it is performed, we will start with the simple model introduced in Chapters 4 and 8. In this model, the free energy $G(x)$ is treated as an effective potential governing the dynamics along x, which is further assumed to obey the Langevin equation introduced in Chapter 4:

$$m\ddot{x} = -G'(x) - \eta\dot{x} + R(t). \tag{9.1}$$

According to Eq. 9.1, the molecule is subjected to three types of forces. The deterministic force $-G'(x)$ originates from the effective potential $G(x)$. The presence of this force guarantees that the Langevin model reproduces the equilibrium distribution of the extension, $w(x) \propto \exp[-G(x)/k_BT]$, exactly (see Chapter 4). When the molecule is subjected to an external force f, such a force must be included in the rhs of the Langevin equation as well (*vide infra*). Interactions with the surrounding solvent further result in a viscous drag force, which acts in the direction opposite the velocity $\dot{x}$ and is proportional to its magnitude (with a proportionality coefficient η), and to a random force $R(t)$, whose properties are described in Chapter 4. It should be emphasized that, although Eq. 9.1, by construction, reproduces the equilibrium properties of the variable x, there is no guarantee that it accurately accounts for the temporal behavior of $x(t)$. The failure of such one-dimensional models to explain catch bonds (see Chapter 8) illustrates this point. Its accuracy notwithstanding, Eq. 9.1 is simple and tractable and the important lessons it offers will be found to be model independent later on. To simplify our analysis even further, the overdamped limit will be assumed (see Chapter 4) and so, similarly to Eq. 4.19, the first term in Eq. 9.1 will be set to zero.

The pulling on this molecule is accomplished via a pulling device (such as an optical trap or an atomic force microscope), which – similarly to the previous chapter— will be modeled as a spring of stiffness γ. That is, the variable X controlled by the experimenter (e.g., the position of the optical trap) is coupled to the molecule's extension via a quadratic potential of the form $\gamma(X - x)^2/2$. We will consider the experiment where the control variable $X(t)$ is increased from an initial value of $X(0) = x_1$ to a final value $X(\tau) = x_2$, with the specific function $X(t)$ chosen at will. These values will be assumed such that they satisfy the inequality $G(x_2) - G(x_1) \gg k_BT$. As a result, the molecule is steered toward a highly improbable state that would be practically unreachable without help from the pulling device. The quantity that we are interested in is the work performed on the molecule by the pulling device. This work is calculated by integrating the instantaneous force that pulls on the molecule, $f(t) = \gamma[X(t) - x(t)]$, over the extension x,

$$W_{12} = \int_0^\tau f(t)dx(t) = \int_0^\tau \gamma[X(t) - x(t)]\dot{x}(t)dt. \tag{9.2}$$

Here x is the solution of the Langevin equation modified to include the additional potential energy term due to the pulling device, i.e.,

$$0 = -\gamma(x - X) - G'(x) - \eta\dot{x} + R(t). \tag{9.3}$$

Multiplying this equation by $\dot{x}$, we obtain

$$\gamma(X - x)\dot{x} = G'(x)\dot{x} + \eta\dot{x}^2 - R(t)\dot{x}.$$

Now, integrating this over the duration τ of the pulling experiment, we find

$$W_{12} = G[x(\tau)] - G[x(0)] + \eta \int_0^\tau dt\,\dot{x}^2 - \int_0^\tau dt\,R(t)\dot{x}(t). \qquad (9.4)$$

If the stretching process is very slow, i.e., $\dot{x} \to 0$, the last two terms in Eq. 9.4 vanish and the relationship

$$W_{12} = G[x(\tau)] - G[x(0)]$$

is recovered, showing that the work performed on the molecule is equal to the change of its free energy. A stiff pulling device forces the extension $x(t)$ to be always close to the control variable $X(t)$. In such a case we can further use the approximation

$$x(0) \approx x_1,\, x(\tau) \approx x_2 \qquad (9.5)$$

and so

$$W_{12} \approx G(x_2) - G(x_1).$$

This result has already been given, for an infinitesimal extension $dx = x_2 - x_1 = \dot{x}dt$, in the preceding chapter. When the stretching velocity $\dot{x}$ is finite, however, two additional terms come into play. The term

$$W_{dis} = \eta \int_0^\tau dt\,\dot{x}^2 \qquad (9.6)$$

is the work done against the friction force $-\eta\dot{x}$ and, ultimately, wasted as heat.[2] The last term of Eq. 9.4 is the work W_R performed by the random force $R(t)$, taken with the opposite sign. The combined work $W_{12} + W_R$ performed on the molecule by the pulling device and by the random force $R(t)$ is thus only partially spent to increase the free energy $G(x)$, with the rest, W_{dis}, dissipated into the surroundings. The random force $R(t)$ can either help stretch the molecule and increase its free energy ($W_R > 0$) or oppose stretching ($W_R < 0$). As a result, the total work W_{12} will vary from one measurement to another, but when averaged over multiple measurements, the contribution of the fluctuating force into this work is zero, since we have $\langle R \rangle = 0$ and since $R(t)$ is statistically independent of $x(t)$. Because the dissipated energy W_{dis} is always nonnegative, the average work performed to stretch the molecule must satisfy the following inequality:

$$\langle W_{12} \rangle \geq G(x_2) - G(x_1). \qquad (9.7)$$

The average work required to stretch the molecule is, therefore, greater than the free energy difference between its final and initial states, unless the process is performed infinitely slowly such that the energy loss vanishes.[3]

---

[2] It should be emphasized that a negligible value of W_{dis} would not imply zero heat exchange with the molecule's environment. As shown in Chapter 8, the free energy change, $G(x_2) - G(x_1)$, necessarily includes heat given off to or absorbed from the surroundings. The Langevin model, however, does not explicitly account for this type of heat exchange because it treats $G(x)$ simply as a potential energy.

[3] If the velocity $\dot{x}$ vanishes, the duration of the process τ becomes infinite. However it is easy to see that W_{dis} defined by Eq. 9.6 will still vanish in this limit. For example, if stretching is performed at a constant velocity $\dot{x} = (x_2 - x_1)/\tau$, then one finds $W_{dis} = \eta(x_2 - x_1)^2/\tau \to 0$.

Consider now the reverse process, in which the molecule starts out in the stretched, high free energy state with $x = x_2$ and relaxes back to the state $x = x_1$. This process can be accomplished at no energy cost by simply letting the molecule relax spontaneously. The elastic (free) energy stored in the stretched molecule can, however, be used to perform mechanical work to pull on the pulling device or—more usefully— on some molecular "cargo." If f is the force exerted on the molecule then $-f$ is the force exerted by the molecule on its cargo or the pulling device. The total work performed by the molecule is given by the same expression as Eq. 9.2 but taken with the opposite sign:

$$W_{21}^* = -\int_0^\tau f(t)dx(t).$$

Here the asterisk is used to remind ourselves that this work is done *by* the molecule (and so it is synonymous with the minus sign in front of the above integral). Repeating the above steps for the contraction process, we find:

$$W_{21}^* = -G[x(\tau)] + G[x(0)] - \eta \int_0^\tau dt\,\dot{x}^2 + \int_0^\tau dt\,R(t)\dot{x}(t).$$

With the assumption of Eq. 9.5, this gives, for the average work performed by the molecule,

$$\langle W_{21}^* \rangle \le G(x_2) - G(x_1). \tag{9.8}$$

Thus the average useful work that can be extracted from the stretched molecule is always less than the stored free energy, except in a slow, lossless process where the two quantities become equal.

Equations 9.7 and 9.8 are well known in thermodynamics. They reflect the fact that most processes in Nature are irreversible and are accompanied by energy dissipation. When a molecule is stretched, some of the work is wasted to overcome friction and so the total work performed exceeds, on the average, the free energy difference between the stretched and the relaxed states. Likewise, if the contraction of a stretched molecule is used to perform useful work, some of the (free) energy stored in the molecule is wasted and so the amount of useful work is less than its maximum possible value. The amount of work $\langle W_{12} \rangle$ ($\langle W_{21}^* \rangle$) approaches the limit of $G(x_2) - G(x_1)$ from above (below) if the process is performed infinitely slowly such that the system is in equilibrium with its surroundings at every point along the way. In this limit, stretching becomes the reverse of contraction. Processes of this type are known in thermodynamics as reversible. The situation discussed above is illustrated in Fig.9.1: If the force f exerted by the pulling device is measured as a function of the extension during a stretching/relaxation cycle, a hysteresis is observed when the cycle is performed at a finite speed, with the work done on the molecule during stretching, W_{12}, typically greater than the work done by the molecule, W_{21}^*, during relaxation. The hysteresis disappears when the experiment is performed increasingly slower. Note that, although the Langevin equation model was used here to derive Eqs. 9.7 and 9.8, these equations remain valid regardless of the dynamical properties of the variable x. A proof of this will be given below.

Eqs. 9.7 and 9.8 are further closely connected to the second law of thermodynamics. This law determines the direction of the "time arrow" of all the processes that occur

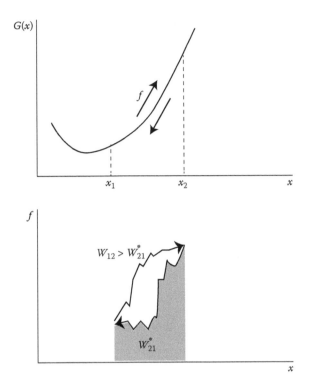

FIGURE 9.1 In a stretching-relaxation cycle carried out on an individual molecule, the typical work W_{12} expended during stretching (where the free energy of the molecule is increased) exceeds the work W_{21}^* done by the molecule during relaxation.

spontaneously. The existence of such a preferred direction of time is, of course, evident in our everyday experience: hot objects, for example, spontaneously get colder (but not hotter), glassware falling on the ground gets irreversibly broken, and so on. The second law sets a fundamental limitation on one's ability to convert energy of random thermal motion into useful work—think, for example, of the ridiculous process where pieces of broken glassware lying on the floor draw thermal energy from the surroundings to spontaneously assemble into a wine glass which proceeds to jump up into your hands, reversing the broken glass mishap.

In one of its traditional formulations, the second law declares that the entropy S of an isolated system (see previous chapter and Appendix B) must increase or stay the same but never decrease on its own, i.e., $\Delta S \geq 0$. Eqs. 9.7 and 9.8 express the second law in the case where the system of interest is not isolated but can exchange energy with its surroundings, assuming that the temperature of the surroundings is maintained at a constant value T. In particular, these equations imply that the free energy of a molecule that it left entirely alone (i.e., no external forces are applied) can only decrease, $G(x_2) < G(x_1)$, which, according to Eq. 8.15, is the same as the second law's claim $S(x_2) > S(x_1)$ if the energy of the molecule does not change, as in the freely jointed chain model discussed in Chapter 8. That is, a highly stretched molecule will tend to spontaneously contract. The molecule's trend to decrease its free energy can, however, be reversed by providing an energy flow into the molecule by, e.g., performing mechanical work. The inequality of Eqs. 9.7 then sets a fundamental limit on the minimum amount of work required to reverse the direction of a spontaneous process.

Before concluding this section and in preparation for the next, it is beneficial to introduce a slightly different definition of work that is obtained by integrating the force-extension curve when the force is reported as a function of the control variable X rather than the molecule's actual extension x:

$$\tilde{W}_{12} = \int_0^\tau f(t)dX(t) = \int_0^\tau \gamma(X - x)\dot{X}(t)dt. \tag{9.9}$$

The two definitions of work are closely related. Indeed, subtraction of one from the other yields

$$\tilde{W}_{12} - W_{12} = \int_0^\tau \gamma(X - x)(\dot{X} - \dot{x})dt = (1/2)\int_0^\tau (d/dt)\gamma(X - x)^2 dt$$

$$= \gamma[X(\tau) - x(\tau)]^2/2 - \gamma[X(0) - x(0)]^2/2. \tag{9.10}$$

The new definition of work labeled with a tilde will be more convenient in the following discussion.

9.2 EXACT RELATIONSHIPS BETWEEN FREE ENERGY AND NONEQUILIBRIUM WORK

The inequalities stated by Eqs. 9.7 and 9.8 apply to molecules as well as to macroscopic bodies. In fact, they have been discovered before the very existence of molecules was firmly established. Recent years have brought the discovery of a new class of *identities* involving the work done in a nonequilibrium process and the corresponding free energy change. While applicable to systems of any size in principle, in practice experimental validation of those identities requires measurements performed on microscopic systems (such as individual molecules). The first of those identities, discovered in the mid 1990's by Christopher Jarzynski [1], has led to a still ongoing flurry of activity both in the theoretical field of nonequilibrium statistical mechanics and in experimental efforts to validate the new statistical mechanics laws. Here, two of those results will be discussed, starting with Jarzynski's original identity.

As discussed before, the precise equations governing the motion of a molecule (and, particularly, the time evolution of the mechanical coordinate x) in a condensed phase environment are generally unknown beyond approximate or *ad hoc* models. If so, the existence of any exact formulas satisfied irrespective of the (unknown) dynamical laws obeyed by $x(t)$ sounds too good to be true. To understand the fundamental origin of such identities as well as to liberate ourselves from any dependence on a particular model (such as the Langevin equation used above), it is beneficial to recall that, no matter how complicated the time evolution of the trajectory $x(t)$ may appear, it, fundamentally, originates from conservative dynamics of a much larger "supersystem," which the molecule of interest is part of. This idea was already introduced in

Chapter 4, where we have argued that energy dissipation is the consequence of an incomplete description of the system. For example, while excess energy stored in a molecule is typically lost, through friction, to the molecules of the solvent, the total energy of the molecule and its solvent may, to a good degree of approximation, be conserved. Although to ensure exact energy conservation the entire Universe may need to be taken into account, in practice, viewing, say, all the degrees of freedom of the molecule of interest plus a sufficiently large set of solvent molecules as an isolated system with conserved energy is often an adequate approximation. At any rate, our findings will turn out to be independent of the precise properties or size of the supersystem, but closely linked to energy conservation and the time reversal symmetry it necessarily obeys.

Let the components of the vectors $\mathbf{r}$ and $\mathbf{p}$ be the coordinates and the momenta of each particle in our supersystem (which include the degrees of freedom of the molecule that we wish to disturb with our pulling device). In the absence of any external forces acting on the supersystem, its total energy,

$$E_0[\mathbf{r}(t), \mathbf{p}(t)],$$

is independent of the time t. This is no longer the case after the experimenter applies a time-dependent perturbation described by the potential

$$V(X(t), \mathbf{r}) = \frac{1}{2}\gamma[x(\mathbf{r}) - X(t)]^2,$$

where $X(t)$ is, as before, the control variable (e.g., the position of the optical trap) that is varied with the goal to increase the molecule's extension x. More generally, our analysis would apply to any collective variable $x = x(\mathbf{r})$ that depends on the supersystem's instantaneous configuration. As a result of explicit time dependence of the perturbation potential $V[X(t), \mathbf{r}]$, the total energy of the supersystem,

$$E[\mathbf{r}(t), \mathbf{p}(t); t] = E_0[\mathbf{r}(t), \mathbf{p}(t)] + V[X(t), \mathbf{r}],$$

now evolves with time according to the equation:

$$dE/dt = \partial V/\partial t = \gamma[X(t) - x(\mathbf{r})]\dot{X}. \tag{9.11}$$

Careful proofs of Eq. 9.11 can be found in just about any classical mechanics text; see, e.g., [2] (see below for an explicit albeit less general derivation). Now if, as before, the initial value of X at the beginning of the stretching experiment is $X(0) = x_1$ and the final value is $X(\tau) = x_2$, then integration of the above equation gives

$$E[\mathbf{r}(\tau), \mathbf{p}(\tau); \tau] - E[\mathbf{r}(0), \mathbf{p}(0); 0] = E_0[\mathbf{r}(\tau), \mathbf{p}(\tau)] - E_0[\mathbf{r}(0), \mathbf{p}(0)]$$
$$+ V[x_2, \mathbf{r}(\tau)] - V[x_1, \mathbf{r}(0)] = \tilde{W}_{12}, \tag{9.12}$$

where the definition of the work given by Eq. 9.9 is used. Equation 9.12 states that the work done by the external perturbation equals the change in the (super) system's energy—a statement of energy conservation that seems almost obvious. Still, one must be careful: Why does the second definition of the work, rather than the first one is used? To illustrate why $\tilde{W}_{12}$ and not W_{12} must appear in Eq. 9.12, consider a simpler problem where the supersystem is a one-dimensional particle whose energy, in the absence of the mechanical perturbation, is $E_0(x, p) = p^2/(2m) + V_0(x)$ with some potential V_0. Newton's second law written for this system in the presence of the pulling device reads:

$$m\ddot{x} = -dV_0/dx - \gamma[x - X(t)].$$

Multiplying this by $\dot{x}$ we obtain

$$m\ddot{x}\dot{x} + (dV_0/dx)\dot{x} = (d/dt)[m\dot{x}^2/2 + V_0(x)] = -\gamma[x - X(t)]\dot{x},$$

which, upon integration over time yields:

$$m\dot{x}^2(\tau)/2 + V_0[x(\tau)] - m\dot{x}^2(0)/2 - V_0[x(0)] = W_{12},$$

with the "old" definition of work used for W_{12}. Now using Eq. 9.10 to switch to $\tilde{W}_{12}$, we find

$$\tilde{W}_{12} = m\dot{x}^2(\tau)/2 + V_0[x(\tau)] + \gamma[x(\tau) - X(\tau)]^2/2 - m\dot{x}^2(0)/2 - V_0[x(0)]$$
$$- \gamma[x(0) - X(0)]^2/2,$$

which agrees with Eq. 9.12.

Before we proceed, the description of the pulling experiment requires more precise specification of the initial state. It will be assumed that, prior to the beginning of the experiment at $t = 0$, the pulling device has already been attached to the molecule with the parameter X set at its initial value $X = x_1$. Moreover, it has been attached long enough that thermal equilibrium has been established and so the probability distribution of the initial coordinates $\mathbf{r}(0)$ and momenta $\mathbf{p}(0)$ of our supersystem is proportional to

$$\exp\left(-\frac{E_1}{k_B T}\right), \tag{9.13}$$

where

$$E_1[\mathbf{r}(0), \mathbf{p}(0)] = E[\mathbf{r}(0), \mathbf{p}(0); 0] = E_0[\mathbf{r}(0), \mathbf{p}(0)] + V[x_1, \mathbf{r}[0]].$$

Starting at $t = 0$, one begins to increase the distance $X(t)$ until it attains its final value $X(\tau) = x_2$. The work $\tilde{W}_{12}$ is then recorded by integrating the curve of the force f vs X. The outcome of the experiment now depends on the initial conditions $\mathbf{r}(0), \mathbf{p}(0)$ drawn from the distribution specified by Eq. 9.13. If, for example, the mean value of the work, $\langle \tilde{W}_{12} \rangle$ is desired, this can, in principle, be obtained by averaging Eq. 9.12 with the probability density defined by Eq. 9.13. We will, however, focus on the average of a different quantity, which is an exponential function of the work, $\exp[-\tilde{W}_{12}/k_B T]$. This can be written as

$$\left\langle \exp\left[-\frac{\tilde{W}_{12}}{k_B T}\right] \right\rangle = q_1^{-1} \int d\mathbf{r}(0) d\mathbf{p}(0) \exp\left(-\frac{\tilde{W}_{12}}{k_B T}\right) \exp\left(\frac{-E_1[\mathbf{r}(0), \mathbf{p}(0)]}{k_B T}\right),$$

where

$$q_1 = \int d\mathbf{r}(0)d\mathbf{p}(0) \exp\left(-\frac{E_1[\mathbf{r}(0), \mathbf{p}(0)]}{k_BT}\right) \tag{9.14}$$

is the factor that ensures that the probability distribution of Eq. 9.13 is normalized. Now, using Eq. 9.12, we have

$$E_1[\mathbf{r}(0), \mathbf{p}(0)] + \tilde{W}_{12} = E_2[\mathbf{r}(\tau), \mathbf{p}(\tau)] = E_0[\mathbf{r}(\tau), \mathbf{p}(\tau)] + V[x_2, \mathbf{r}(\tau)]], \tag{9.15}$$

which gives

$$\left\langle \exp\left[-\frac{\tilde{W}_{12}}{k_BT}\right]\right\rangle = \int d\mathbf{r}(0)d\mathbf{p}(0) \exp\left(\frac{-E_2[\mathbf{r}(\tau), \mathbf{p}(\tau)]}{k_BT}\right)\Big/q_1$$

$$= \int d\mathbf{r}(\tau)d\mathbf{p}(\tau) \exp\left(\frac{-E_2[\mathbf{r}(\tau), \mathbf{p}(\tau)]}{k_BT}\right)\Big/q_1, \tag{9.16}$$

where conservation of phase-space volume along a classical trajectory (i.e., the Liouville theorem, see Chapter 5) has been noted, which allowed us to replace the infinitesimal volume $d\mathbf{r}(0)d\mathbf{p}(0)$ by $d\mathbf{r}(\tau)d\mathbf{p}(\tau)$. Finally, if we define

$$q_2 = \int d\mathbf{r}d\mathbf{p}\exp\left[-\frac{E_2(\mathbf{r}, \mathbf{p})}{k_BT}\right], \tag{9.17}$$

we arrive at the remarkably simple equation:

$$\left\langle \exp\left[-\frac{\tilde{W}_{12}}{k_BT}\right]\right\rangle = q_2/q_1. \tag{9.18}$$

Moreover, since q_1 and q_2 have the form of partition functions (cf. Chapters 4–5 and Appendix B), we can define associated free energies through

$$q_{1,2} = e^{-\frac{G_{1,2}}{k_BT}},$$

which gives us the original formula first derived by Jarzynski [1]:

$$\left\langle \exp\left[-\frac{\tilde{W}_{12}}{k_BT}\right]\right\rangle = \exp\left[-\frac{G_2 - G_1}{k_BT}\right]. \tag{9.19}$$

It is tempting at this point to declare Eq. 9.19 to be the exact equality connecting the quantities appearing in Eq. 9.7. This would be true if $G_{1,2}$ were the same as the free energies $G(x_{1,2})$ appearing in Eq. 9.7, but, unfortunately, this is not the case. In particular, as seen from their definition, the free energies $G_{1,2}$ depend on the stiffness γ of the pulling device and, therefore, are not intrinsic properties of the molecule alone. Recovering $G(x_{1,2})$ from Jarzynski's identity thus turns out to be a nontrivial challenge, which was taken up by several authors [3,4]. Here, the discussion will be limited to the case of pulling via a stiff spring, where the analysis becomes much simpler. Rewriting the definition of the partition functions as

$$q_{1,2} = \int d\mathbf{r}d\mathbf{p}\exp\left[-\frac{E_0(\mathbf{r}, \mathbf{p})}{k_BT}\right]\exp\left[-\frac{\gamma(x_{1,2} - x)^2}{2k_BT}\right],$$

we note that they are explicit functions of x_1 and x_2, and—additionally—of the spring constant γ. If one lowers the temperature or increases γ, the contributions from molecular configurations with the extension x significantly different from $x_{1,2}$ become progressively suppressed. This invites the approximation

$$\exp\left[-\frac{\gamma(x_{1,2} - x)^2}{2k_B T}\right] \approx A\delta(x_{1,2} - x),$$

where A is a normalization factor, and results in

$$q_{1,2} = A \int d\mathbf{r} d\mathbf{p} \exp\left[-\frac{E_0(\mathbf{r}, \mathbf{p})}{k_B T}\right] \delta(x_{1,2} - x). \tag{9.20}$$

Physically, Eq. 9.20 instructs us to sum (integrate) over all possible configurations weighted according to the canonical distribution corresponding to E_0, but pick only those configurations that satisfy the condition $x_{1,2} = x$. The result, to within an unimportant normalization factor, is just the equilibrium probability distribution of $x_{1,2}$ for the unperturbed system (i.e., the one described by the energy E_0),

$$q_{1,2} \propto w(x_{1,2}) = e^{-\frac{G(x_{1,2})}{k_B T}}.$$

We now arrive at the approximate version of Jarzynskii's formula that relates the equilibrium free energy $G(x)$ to the work done in a nonequilibrium pulling experiment:

$$\left\langle \exp\left[-\frac{\tilde{W}_{12}}{k_B T}\right] \right\rangle = \exp\left[-\frac{G(x_2) - G(x_1)}{k_B T}\right]. \tag{9.21}$$

The inequalities 9.7 and 9.8 can be viewed as direct consequences of Eq. 9.21. To show this, we use the inequality

$$\langle e^a \rangle \geq e^{\langle a \rangle} \tag{9.22}$$

satisfied by any random variable a.[4] Setting $a = \tilde{W}_{12}$, we obtain

$$\exp\left[-\frac{\langle \tilde{W}_{12} \rangle}{k_B T}\right] \leq \left\langle \exp\left[-\frac{\tilde{W}_{12}}{k_B T}\right] \right\rangle = \exp\left[-\frac{G(x_2) - G(x_1)}{k_B T}\right],$$

and, consequently,

$$-\frac{\langle \tilde{W}_{12} \rangle}{k_B T} \leq -\frac{G(x_2) - G(x_1)}{k_B T}. \tag{9.23}$$

Given our earlier assumption that $G(x_2) > G(x_1)$, the above inequality is equivalent to

$$\langle \tilde{W}_{12} \rangle \geq G(x_2) - G(x_1), \tag{9.24}$$

which is analogous to Eq. 9.7 (except for the slightly different definition of work). Notice now that none of the arguments used to derive Eq. 9.21 required any particular

---

[4] To prove this inequality, notice that $e^b \geq 1 + b$ for any number b, where the equality holds only at $b = 0$. Setting $b = a - \langle a \rangle$ and averaging over the distribution of a immediately yields Eq. 9.22.

relationship between the initial and the final free energies and so it must hold when the free energy of the final state is lower than that of the initial state. With the same assumption that $G(x_2)$ is higher than $G(x_1)$, let us then apply our result to the reverse process, $2 \rightarrow 1$, where the molecule is allowed to relax into a lower free energy state. Simply swapping the initial and final indices in Eq. 9.24, we obtain

$$\frac{\langle \tilde{W}_{21}^* \rangle}{k_B T} = -\frac{\langle \tilde{W}_{21} \rangle}{k_B T} \leq -\frac{G(x_1) - G(x_2)}{k_B T},$$

where the asterisk, as before, indicates that the computed work is that performed *by* the molecule on the pulling device. This is equivalent to

$$\langle \tilde{W}_{21}^* \rangle \leq G(x_2) - G(x_1),$$

which is analogous to Eq. 9.8.

The deviations of the work from its average are even more interesting than the average itself. Let

$$\Delta \tilde{W}_{12} = \tilde{W}_{12} - [G(x_2) - G(x_1)]$$

be the excess work spent to stretch the molecule, relative to that required in the reversible limit. On the average, this difference must be nonnegative,

$$\langle \Delta \tilde{W}_{12} \rangle \geq 0.$$

However, Jarzynski's identity, Eq. 9.21, written in terms of the excess work, reads

$$\left\langle \exp\left[-\frac{\Delta \tilde{W}_{12}}{k_B T} \right] \right\rangle = 1, \tag{9.25}$$

implying that $\Delta \tilde{W}_{12}$ cannot be always positive. Indeed, positive values of the excess work would result in the lhs of Eq. 9.25 being less than one. Thus the rare single-molecule trajectories for which $\Delta \tilde{W}_{12} < 0$ and which, therefore, violate the second law of thermodynamics are the key to satisfying the Jarzynski identity. Such violation is acceptable for a microscopically small system, since fluctuating forces imposed on such a system by the surrounding molecules may conspire to help increase its free energy, reducing the required work (cf. the term W_R in the previous section). Such fluctuations will grow increasingly rare when the system's size grows. No matter how rare, however, they still contribute to the average of Eq. 9.25 because they are exponentially amplified! This observation, in particular, implies that small systems, such as individual molecules, are most likely candidates for an experimental test of Eq. 9.21 because work fluctuations could be readily observable in them. Such tests have, indeed, been performed in single-molecule studies of, e.g., RNA molecules pulled by optical tweezers [5], confirming Jarzynski's identity and demonstrating that equilibrium free energy differences can be estimated from nonequilibrium work distributions through Eq. 9.21. Of course, being a rigorously derivable identity, Jarzynski's result does not really need an experimental proof and its failure to account for experimental observations would most likely indicate a problem with the experiment, not theory. Yet agreement between experiment and theory is always gratifying.

Whatever the distribution of the work $\tilde{W}_{12}$ is, it must satisfy the constraint imposed by Eq. 9.21. Gavin Crooks discovered a further identity relating the distributions of work measured in the stretching and relaxation processes [6], assuming that the time dependence of the control variable $X(t)$ during relaxation is the exact time-reversal of that during stretching. For example, if X is increased at a constant velocity during stretching, it must be decreased at the same velocity during relaxation. Let $w_{12}(\tilde{W})$ be the distribution of the work performed on the molecule in the stretching process and $w_{12}^*(\tilde{W})$ the distribution of the work performed by the molecule in the relaxation process. As before, the asterisk is there to remind us that the work measured during relaxation is taken with the opposite sign. To avoid cumbersome notation, this asterisk, as well as the subscript "12" or "21" indicating the direction of the process, is now associated with the distributions w while the work measured in each case is simply denoted $\tilde{W}$. The Jarzynski identity can now be written as

$$\int_{-\infty}^{\infty} \exp\left[-\frac{\tilde{W}}{k_B T}\right] w_{12}(\tilde{W}) d\tilde{W} = \exp\left[-\frac{G(x_2) - G(x_1)}{k_B T}\right],$$

while Eq. 9.24 amounts to

$$\int_{-\infty}^{\infty} \tilde{W} w_{12}(\tilde{W}) d\tilde{W} \leq G(x_2) - G(x_1).$$

What Crooks showed is that the distributions of work done in the direct $(1 \to 2)$ and reverse $(2 \to 1)$ processes must always satisfy the identity:

$$w_{12}(\tilde{W}) = \exp\left[\frac{\tilde{W} - G_2 + G_1}{k_B T}\right] w_{21}^*(\tilde{W}), \qquad (9.26)$$

where, as before, $G_{1,2}$ can be replaced by $G(x_{1,2})$ if the pulling spring is sufficiently stiff. As an immediate consequence of the Crooks equation, one observes that $w_{12}(\tilde{W}) = w_{21}^*(\tilde{W})$ if the work $\tilde{W}$ equals the equilibrium work $G_2 - G_1$. This offers an alternative method of estimating the free energy difference from a nonequilibrium work distribution. This method is illustrated in Figure 9.2: If the stretching/relaxation experiments are carried out away from the equilibrium conditions, the obtained histogram of the work to stretch the molecule will be shifted to the right of the equilibrium work value $G_2 - G_1$. Likewise, the distribution of the work done by the molecule during its contraction will be shifted to the left of the equilibrium value. The location of the point $\tilde{W}$ where these two histograms cross gives exactly the equilibrium work.

The Jarzynski identity can be immediately obtained from the Crooks equation as follows: Both work distributions must be, of course, normalized. Multiplying both sides of Eq. 9.26 by $\exp[-\frac{\tilde{W}}{k_B T}]$ and integrating the result, we obtain:

$$\int_{-\infty}^{\infty} \exp\left[-\frac{\tilde{W}}{k_B T}\right] w_{12}(\tilde{W}) d\tilde{W} = \left\langle \exp\left[-\frac{\tilde{W}}{k_B T}\right] \right\rangle$$

$$= \int_{-\infty}^{\infty} \exp\left[-\frac{G_2 - G_1}{k_B T}\right] w_{21}^*(\tilde{W}) d(\tilde{W})$$

$$= \exp\left[-\frac{G_2 - G_1}{k_B T}\right],$$

which is Jarzynski's result.

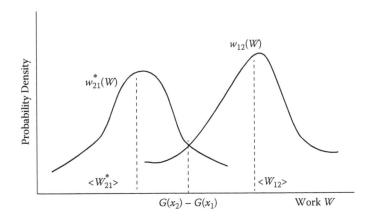

FIGURE 9.2 Away from equilibrium conditions, the probability distribution of the work done on the molecule during stretching is shifted to the right relative to the free energy cost of the process, $\Delta G = G(x_2) - G(x_1)$, while the probability distribution of the work done by the molecule in the course of its contraction is shifted to the left of ΔG. According to the Crooks equation, the tails of these two distributions must intersect exactly at ΔG thus providing an estimate of the equilibrium work.

9.3 ENERGY DISSIPATION IN BIOLOGICAL MOLECULES: SACRIFICIAL BONDS AND MOLECULAR SHOCK ABSORBERS

Energy dissipation, both at the molecular and macroscopic scales, is a process crucial to our well-being. Much of research is currently being invested in minimizing the energy wasted by automobile engines. The efficiency of various molecular machines found in living organisms and converting chemical energy into mechanical work may, likewise, impact their survival. Those machines, performing myriads of chores in the cell, from muscle contraction to DNA manipulation, will be the subject of further discussion in Chapter 10.

Energy dissipation, however, is not always bad. Without efficient energy dissipation provided by shock absorbers, for example, a car would be bouncing up and down every time it hits a pothole, resulting in an unpleasant and dangerous ride. Certain biological molecules have a similar purpose to act as microscopic shock absorbers. They can dissipate energy in quantities much greater than expected from the viscous friction mechanism discussed earlier. An example is offered by the muscle protein titin, a long molecular spring which is believed to prevent microscopic damage to the muscle tissue by absorbing energy when, e.g., the muscle is overstretched. Titin is a "polyprotein," i.e., a molecule composed of multiple sequentially connected globular domains. Titin (and, particularly, one of its domains known as "I27") has been the focus of many single-molecule pulling studies, which uncovered how such a modular design can lead to efficient energy dissipation. Those studies usually employed the atomic force microscope (AFM) to pull on polyprotein constructs consisting of multiple I27s (Figure 9.3). As the molecule is stretched, its tension f (which can be measured through the deflection of the AFM cantilever) is recorded as a function

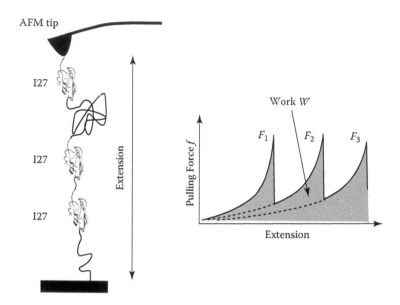

FIGURE 9.3 In AFM pulling studies of titin, polyprotein constructs involving tandem I27 domain repeats are stretched. The recorded force-extension curves display multiple force peaks resulting from the unfolding of individual domains.

of the molecule's overall extension. A typical force-extension curve exhibits a series of peaks, each followed by an abrupt drop in the force (Figure 9.3). Such drops result from the unfolding of individual domains in the chain. Remarkably, the work performed to stretch the polyprotein, measured as the area under the force-extension curve, is typically found to exceed the equilibrium free energy cost of its unfolding by several orders of magnitude.

To understand the origin of this behavior, let us focus on the unfolding of just one I27 domain. This domain is incorporated within a chain containing other domains, some of which are already unfolded. An unfolded protein is random and we could model it using the freely jointed chain (FJC) model from the previous chapter (cf. Fig. 8.2). At sufficiently low forces/extensions its "entropic" elasticity can be approximated by Eq. 8.8, which effectively replaces the unfolded chain with a Hookean spring, whose stiffness is equal to

$$\gamma_0 = \frac{3k_BT}{n_0l^2},$$

where n_0 is the number of the FJC "rods" and l is the length of a single rod.[5] In contrast, the folded domain is compact and exhibits a much higher stiffness. Moreover, the ends of the folded I27 are close to one another and so the extension of the folded

---

[5] The linear spring approximation breaks down at high extensions and is often inadequate if a quantitative description of titin pulling is desired. Results from Section 8.5 offer a better approximation in such cases but will not be used, as only qualitative insight is sought here.

domain is negligible when compared to that of the unfolded chain. Thus, the overall spring constant of the entire chain can still be approximated by γ_0. For the purpose of calculating the mechanical response of such a chain, the folded chain segment is essentially equivalent to a bond, or a crosslink, that creates a loop of length Δn, equal to the length of the I27, shielding it from the stretching force (Figure 9.4). When the I27 domain becomes unfolded, this loop is liberated, resulting in additional slack within the chain. As a result, the total spring constant of the chain drops to

$$\gamma_1 = \frac{3k_B T}{(n_0 + \Delta n)l^2}.$$

Given an extension x_u measured at the moment the domain unfolds, this results in a drop of the force from $f_0 = \gamma_0 x_u$ to $f_1 = \gamma_1 x_u$ (Fig. 9.4). The excess work expended to pull on the stiffer chain containing the folded domain, relative to the work that would be required to stretch the chain with the I27 already unfolded, equals the gray area shown in (Fig. 9.4) and can be estimated as

$$\Delta W = \int_0^{x_u} (\gamma_0 x')dx' - \int_0^{x_u} (\gamma_1 x')dx' = \frac{f_0^2}{2\gamma_0} - \frac{f_0^2 \gamma_1}{2\gamma_0^2} = \frac{f_0^2 l^2}{6 k_B T} \frac{1}{n_0^{-1} + \Delta n^{-1}}. \quad (9.27)$$

The I27 domain consists of $\Delta n + 1 = 98$ repeat units (amino acid residues), each approximately of length $l = 3.8\text{Å}$.[6] The value of n_0 depends on exactly how I27 is attached to the AFM setup and on how many I27 domains are already unfolded. To be specific, let us use the value equal to the length of one domain, $n_0 = \Delta n = 97$. Using a typical experimental value of $f_0 = 200$ piconewtons (pN) for the unfolding force, Eq. 9.27 now gives:

$$\Delta W \approx 2700 k_B T.$$

Here the extra work required to unfold the I27 domain is expressed in units of thermal energy $k_B T$. It is instructive to compare this quantity to the free energy difference $\Delta G_u = G_u - G_f$ between the unfolded and folded states of the same protein. This difference can be measured by studying the thermodynamic properties of the same protein in solution and is related to the measurable probabilities $w_{u,f}$ of finding I27 in its unfolded and folded states in the absence of the force:

$$\frac{w_u}{w_f} = \frac{q_u}{q_f} = \frac{e^{-\beta G_u}}{e^{-\beta G_f}} = e^{-\frac{\Delta G_u}{k_B T}},$$

where $q_{u,f}$ are the corresponding partition functions (cf. Appendix B). Then we have

$$\Delta G_u = -k_B T \ln(w_u/w_f).$$

For I27, this quantity was measured to be of order of $\Delta G \approx 10 k_B T$, more than two orders of magnitude less than the work required to unfold the same domain mechanically!

---

[6] The assumption that the directions of adjacent units of our polymer chain are uncorrelated, as in a freely jointed chain, is not very accurate. It is, however, possible to view l as the *effective* length l over which the

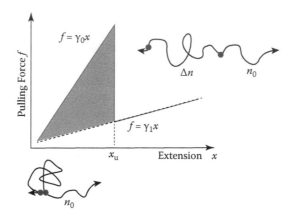

FIGURE 9.4 A simple model of domain unfolding replaces the folded domain with a "sacrificial bond," which shields the domain's polypeptide chain (of length Δn) from the stretching force. The unfolded segments of the chain are modeled as freely jointed chains, which are further approximated (see Chapter 8) as linear springs with a stiffness that is inversely proportional to the chain contour length. When the sacrificial bond is finally broken at some extension value $x = x_u$, the hidden length Δn is released and the spring constant—and the corresponding force—drop to a lower value. The extra work required to rupture the bond through this process is shown as the gray area.

Equation 9.27 further shows that the shock-absorbing efficiency of a protein domain can be improved in two ways: either increase the length Δn of the polypeptide chain forming the domain or increase the peak force f_0. But what determines f_0? Clearly, it must depend on the domain structure and the cohesive interactions that have led to its formation. Since the thermodynamic stability of the folded domain is determined by the same interactions, it is not unreasonable to expect some correlation between f_0 and ΔG_u. However, experimental studies of various protein domains have revealed that more thermodynamically stable domains (i.e., ones with higher values of ΔG_u) do not necessarily unfold at higher forces. Two complementary arguments help explain this finding. First, based on our everyday experience, the energetic cost of overcoming cohesive interactions that keep objects together (or, in other words, the cost of breaking those objects) cannot be the whole story. Trying, for example, to break a matchstick by pulling its ends apart should immediately convince us that the required force strongly depends on how it is applied. In the context of proteins with their atoms arranged in complex geometric shapes, this implies that the direction of the pulling force in relation to the domain geometry will strongly affect its resistance to unfolding by force. Second, the kinetics of unfolding is controlled not by the free energy difference between the unfolded and folded states but by the free energy

---

directional memory is lost. Crudely speaking, at length scales much shorter than l the polymer behaves as a rigid rod while at length scales much longer than l it looks more like dental floss. This effective length was first introduced by the Swiss physical chemist Werner Kuhn and is known as the Kuhn length. For polypeptides, the Kuhn length is somewhat longer than the length of a single structural unit, resulting in a value of ΔW that is somewhat larger than the one estimated here.

barrier between the two. To illustrate this point, let us recall the free energy profile depicted in Fig. 8.9 from Chapter 8. Except at very high force, the domain's free energy, as a function of its extension, has two minima, the left one corresponding to the folded state and the right one to the unfolded one. When the force f is increased, the right (unfolded) minimum moves down relative to the left one and, eventually, the unfolded state becomes more likely, thermodynamically, than the folded one. However, it takes time to get the more favorable state and this time is largely controlled by the barrier between the two minima. If the force is increased too quickly, the domain may remain trapped in its folded state, still separated from the lower free energy state by the barrier. This is precisely what happens in the studies of I27, where kinetics happens to rule over equilibrium thermodynamics. In order to pay the minimal thermodynamic penalty to unfold this domain, the pulling experiment would have to be performed at an excruciatingly slow rate—a proposition unacceptable to a graduate student who prefers to defend the PhD thesis in a finite number of years. More precisely, the pulling time should be much longer than the time it takes to cross the barrier once the folded state has become favorable thermodynamically. But because this time is exponentially sensitive to the height of the barrier and because this height itself is exponentially sensitive to the force (cf. Eq. 8.24), not all protein domains behave like I27. In fact, many domains other than I27 unfold under near equilibrium conditions even when studied by impatient graduate students.

Stated differently, the unfolding force depends on the rate or the timescale of stretching, relative to the internal timescale of its unfolding dynamics. This dependence can be further investigated by adopting models developed in the previous chapter. Specifically, the probability of unfolding in the presence of a force f can be approximated using Eq. 8.24, which we will write in the form

$$k(f) = k(0)e^{f/f_c}, \tag{9.28}$$

where the characteristic force $f_c = k_B T / \Delta x$ describes the sensitivity of the unfolding rate to force and can be estimated if the length Δx (equal to the change in the domain extension x when going from the folded state to the transition state) is known (Fig. 8.9). As the entire chain is stretched, the force $f = f(t)$ is increased until an unfolding event takes place at $f = f_0$. The probability that the domain has not unfolded up to a time t, $w_f(t)$,[7] satisfies the differential equation[8]

$$dw_f/dt = -k[f(t)]w_f. \tag{9.29}$$

Integrating this, we find

$$w_f(t) = e^{-\int_0^t k[f(t')]dt'}.$$

---

[7] If the chain contains several folded domains, the probability that *none* of them unfolds satisfies the same differential equation as below but the rate coefficient $k(f)$ must be multiplied by the number of domains
[8] Even if Equation 9.29 can be regarded as exact at constant force, it, strictly speaking, becomes an approximation in the case of a time-dependent force $f(t)$ as the latter may disturb thermal equilibrium in the initial (folded) state. The implicit assumption made in Eq. 9.29 is that the pulling timescale is much slower as compared to the relaxation timescales within the folded ensemble so that the internal degrees of freedom of the folded protein remain in thermal equilibrium at any instant.

The probability that the domain unfolds between t and $t + dt$ is (cf. Chapter 3)

$$w_u(t)dt = -(dw_f/dt)dt = k[f(t)]e^{-\int_0^t k[f(t')]dt'}dt.$$

If, for example, the force is increased linearly at a constant rate,

$$f = at,$$

then, using Eq. 9.28, we find the following equation for the probably distribution of the unfolding time $w_u(t)$:

$$w_u(t) = k(0)\exp\left[\frac{at}{f_c}\right]\exp\left[-\frac{k(0)f_c}{a}(e^{at/f_c} - 1)\right],$$

and the corresponding distribution $w_u(f)$ of the unfolding force $f = at$ is

$$w_u(f) = a^{-1}k(0)\exp\left[\frac{f}{f_c}\right]\exp\left[-\frac{k(0)f_c}{a}(e^{f/f_c} - 1)\right].$$

While the exact value of the unfolding force is random, it should not be too far from the most probable value corresponding to the maximum of $w_u(f)$. We find this value by setting the derivative $dw_u(t)/dt$ to zero, which gives

$$f_0 = f_c \ln \frac{a}{k(0)f_c}.$$

Unlike viscous drag force, which is proportional to the velocity at which an object moves, this force has a much weaker, logarithmic dependence on the rate of loading a (which—if the entire polyprotein chain is still modeled as a linear spring—is proportional to the pulling speed). It further follows that protein domains displaying high mechanical resistance must have a low zero-force unfolding rate coefficient $k(0)$ and/or high value of the characteristic force f_c.

The process where large amounts of energy are dissipated through the breaking of mechanically resistant "bonds" and release of an initially hidden length (Fig. 9.4) was termed the "sacrificial bond" mechanism [7]. On a macroscopic scale, the dissipated energy is related to the material property called toughness. Toughness quantifies the material's ability to absorb energy without breaking. Toughness is different from strength: Brittle materials may be strong yet they display low toughness and fracture easily. The sacrificial bond mechanism discovered through single-molecule AFM studies of proteins provides an explanation of the extraordinary toughness of a variety of Nature made biomaterials such as, e.g., spider silk and bone [7].

9.4 FURTHER DISCUSSION: PROOF OF THE CROOKS IDENTITY

To prove the Crooks equation, we start by writing the work probability distribution $w_{12}(\tilde{W})$ as the average of the delta-function

$$w_{12}(\tilde{W}) = \langle\delta(\tilde{W}_{12} - \tilde{W})\rangle_{E_1} = \langle\delta(E_2[\mathbf{r}(\tau), \mathbf{p}(\tau)] - E_1[\mathbf{r}(0), \mathbf{p}(0)] - \tilde{W})\rangle_{E_1}. \quad (9.30)$$

The subscript E_1 indicates that the average is over the initial conditions, $\mathbf{r}(0)$ and $\mathbf{p}(0)$, obeying the canonical distribution corresponding to the energy function $E_1[\mathbf{r}(0), \mathbf{p}(0)]$. Here, Eq. 9.15 was used to express the pulling work as a difference between the final and initial energies. Writing this average explicitly and performing a series of straightforward manipulations, we find

$$
w_{12}(\tilde{W})
$$

$$
= q_1^{-1} \int d\mathbf{r}(0) d\mathbf{p}(0) \delta\{E_2[\mathbf{r}(\tau), \mathbf{p}(\tau)] - E_1[\mathbf{r}(0), \mathbf{p}(0)] - \tilde{W}\} e^{\frac{-E_1[\mathbf{r}(0), \mathbf{p}(0)]}{k_B T}}
$$

$$
= q_1^{-1} \int d\mathbf{r}(0) d\mathbf{p}(0) \delta\{E_2[\mathbf{r}(\tau), \mathbf{p}(\tau)] - E_1[\mathbf{r}(0), \mathbf{p}(0)] - \tilde{W}\} e^{\frac{-E_2[\mathbf{r}(\tau), \mathbf{p}(\tau)] + \tilde{W}}{k_B T}}
$$

$$
= \frac{q_2}{q_1} e^{\frac{\tilde{W}}{k_B T}} q_2^{-1} \int d\mathbf{r}(0) d\mathbf{p}(0) \delta\{E_2[\mathbf{r}(\tau), \mathbf{p}(\tau)] - E_1[\mathbf{r}(0), \mathbf{p}(0)] - \tilde{W}\} e^{\frac{-E_2[\mathbf{r}(\tau), \mathbf{p}(\tau)]}{k_B T}}
$$

$$
= e^{\frac{-G_2 + G_1 + \tilde{W}}{k_B T}} q_2^{-1} \int d\mathbf{r}(\tau) d\mathbf{p}(\tau) \delta\{E_2[\mathbf{r}(\tau), \mathbf{p}(\tau)] - E_1[\mathbf{r}(0), \mathbf{p}(0)] - \tilde{W}\} e^{\frac{-E_2[\mathbf{r}(\tau), \mathbf{p}(\tau)]}{k_B T}}.
$$

$$(9.31)$$

Here, as in the derivation of Jarzynski's equation, conservation of the phase space volume was exploited. Now we are going to take advantage of the time reversal symmetry obeyed by our supersystem: For any trajectory that starts at $t = 0$ from an initial configuration $\mathbf{r}(0)$ and an initial momentum (more precisely, vector of momenta) $\mathbf{p}(0)$ and ends in configuration $\mathbf{r}(\tau)$ with a momentum $\mathbf{p}(\tau)$, there is an exactly reversed trajectory of the same duration, τ, that starts from the phase-space point $\{\mathbf{r}(\tau), -\mathbf{p}(\tau)\}$ and is terminated at the phase-space point $\{\mathbf{r}(0), -\mathbf{p}(0)\}$, provided that the time evolution of the control variable $X(t)$ (and, therefore, of the time-dependent perturbation introduced by pulling) is exactly reversed. As a consequence, we can swap the initial and the final times and reverse the momenta, leading to

$$
e^{\frac{-G_2 + G_1 + \tilde{W}}{k_B T}} q_2^{-1} \int d\mathbf{r}(0) d\mathbf{p}(0) \delta\{E_2[\mathbf{r}(0), -\mathbf{p}(0)] - E_1[\mathbf{r}(\tau), -\mathbf{p}(\tau)] - \tilde{W}\} e^{\frac{-E_2[\mathbf{r}(0), -\mathbf{p}(0)]}{k_B T}},
$$

without changing the magnitude of Eq. 9.31. Recognizing that energy is a quadratic function of the momenta so that its value does not change if the sign of all the momenta is reversed, and using the fact that the delta function is an even function of its argument, we write this as

$$
e^{\frac{-G_2 + G_1 + \tilde{W}}{k_B T}} g(\tilde{W})
$$

where

$$
g(\tilde{W}) = q_2^{-1} \int d\mathbf{r}(0) d\mathbf{p}(0) \delta\{E_1[\mathbf{r}(\tau), \mathbf{p}(\tau)] - E_2[\mathbf{r}(0), \mathbf{p}(0)] - (-\tilde{W})\} e^{\frac{-E_2[\mathbf{r}(0), \mathbf{p}(0)]}{k_B T}}.
$$

$$(9.32)$$

Comparing our expression for $g(\tilde{W})$ with Eq. 9.30 we notice that they are of exactly the same form if the initial state is 2 and the final state is 1; that is, we are dealing with a relaxation process. Finally, recalling that $-\tilde{W}$ is the work performed by the molecule, we arrive at the Crooks identity, Eq. 9.26. Our derivation makes it obvious that, in order for Eq. 9.26 to be valid, the relaxation experiment must start from thermal equilibrium corresponding to energy E_2 and, once it starts, the time evolution of the the control variable $X(t)$ must be the exact reversal of its time dependence during stretching.

REFERENCES

1. Christopher Jarzynski, "Nonequilibrium equality for free energy differences", *Physical Review Letters*, vol. 78, 2690-2693, 1997.
2. L.D. Landau and E.M. Lifshitz, *Mechanics*, Elsevier, 2005.
3. Gerhard Hummer and Attila Szabo, "Thermodynamics and kinetics from single-molecule force spectroscopy", in Eli Barkai, Frank Brown, Michel Orrit, and Haw Yang (Editors), *Theory and Evaluation of Single-Molecule Signals*, World Scientific, 2008.
4. S. Park, F. Khalili-Araghi, E. Tajkhorshid, and K. Schulten, "Free energy calculation from steered molecular dynamics simulations using Jarzynski's equality", *Journal of Chemical Physics*, vol. 119, p. 3559, 2003
5. J. Liphardt, S. Dumont, S.B. Smith, I. Tinoco, Jr., and C.J. Bustamante, "Information from nonequilibrium measurements in an experimental test of Jarzynski's equality", *Science*, vol. 296, p.1832, 2002.
6. G.E. Crooks, "Path-ensemble averages in systems driven far from equilibrium", *Physical Review E*, vol. 61, p. 2361, 2000.
7. B.L. Smith, T. E. Schäffer, M. Viani, J. B. Thompson, N. A. Frederick, J. Kindt, A. Belcher, G. D. Stucky, D. E. Morse, and P. K. Hansma, "Molecular mechanistic origin of the toughness of natural adhesives, fibres, and composites", *Nature*, vol. 399, p. 761, 1999.

10 Single-Molecule Phenomena in Living Systems

> ... incredibly small groups of atoms, much too small to display exact statistical laws, do play a dominating role in the very orderly and lawful events within a living organism.
>
> Erwin Schrödinger, *What is Life?*

Chemical kinetics and thermodynamics, as traditionally formulated, are well suited for the description of chemical reactors where large quantities of chemicals are uniformly mixed. The quantities of interest are then the amounts of each type of chemical in the mixture. Living organisms are "reactors" that contemporaneously carry out bewilderingly complex networks of chemical reactions. Such complexity is made possible by the compartmentalization of the organism, with different processes occurring at different locations. The basic compartment common to most organisms is the cell, which, in turn, may include smaller compartments such as mitochondria and the cell nucleus. While each such compartment could be viewed as a minireactor, it often contains merely a small, countable number of molecules of the same type (the genomic DNA being one obvious example). As pointed out in Chapter 3, the equations describing bulk chemical kinetics cannot be directly used in such a case. The single-molecule view, in contrast, provides a natural framework for describing cellular phenomena. Single-molecule approaches also offer significant experimental benefits. They, in particular, enable the researchers to monitor molecular phenomena as they unfold directly in the cell and to probe mechanical forces produced by molecular machines.

The common cliche that "equilibrium is death" underscores that life is fundamentally a nonequilibrium phenomenon. A truly nonequilibrium situation would, however, be a likely death to any tractable physical theory purporting to rationalize the behavior of biological systems, as even the most basic concepts of, say, pressure or temperature do not apply to nonequilibrium systems. Should a doctor be ridiculed for taking your temperature on the grounds that the notion of temperature is, technically, inapplicable to a nonequilibrium system such as the human body? Probably not. The fact that our body's temperature is well defined for any practical purposes has microscopic implications. We expect, for example, the velocities of the atoms composing our body to obey the Maxwell-Boltzmann distribution corresponding to this temperature. The existence of the Maxwell-Boltzmann distribution was essential for our derivation of the microscopic expression for a reaction rate (see Chapter 5), which has led to an Arrhenius-type expression for the rate coefficient, $k = A \exp[-E/(k_B T)]$. The validity of this expression is widely (and justifiably) adopted by the scientists who study chemical transformations in living organisms. It is then clear that living systems must be, at least, in a state of *partial equilibrium* [1], where some of the degrees of freedom (e.g., the momenta of all the atoms) display the behavior prescribed by the laws of equilibrium statistical mechanics and thermodynamics. The assumption of partial equilibrium was implicit in Chapter 4 in our discussion of stochastic models of molecular dynamics, such as the Langevin equation or the master equation. Indeed, those models assume that the motion of the molecule itself does not disrupt thermal equilibrium of the surroundings. For example, the statistical properties of the random force $R(t)$ in the Langevin equation are not affected by the molecule's motion and depend only on the temperature, T, of the surroundings.

While thermal equilibrium within a body is often a good assumption, we maintain our own temperature and, therefore, are clearly not in thermal equilibrium with our surroundings. This is a consequence of the energy stored by the food we eat and, ultimately, "burned" by our organism and dissipated as heat. As we eat and breathe, molecules consumed in the metabolic processes within our organism are replenished while the waste products of those processes are removed. It is often a useful approximation to consider the amounts of chemicals within a cell or an organism as a whole to be in a *steady state*. That is, those amounts remain (approximately) constant but not equal to the amounts corresponding to chemical equilibrium, because molecules are constantly added to or removed from the system.

In earlier chapters, the term "nonequilibrium" was used in two distinct contexts. In Chapter 9 we described the behavior of molecules driven away from their typical conformations by mechanical forces. Driving the molecules too fast resulted in nonequilibrium, or irreversible behavior, where, for example, the free energy stored in a system can only be partially used to perform mechanical work, with the rest wasted through dissipative processes. On the other hand, using a simple isomerization reaction A = B as an example in Chapter 3, we showed that equilibrium (or its lack) can be the property of an ensemble but not of its individual members. That is, although the amounts of A and B may take on highly nonequilibrium values, each individual molecule within the ensemble does not "know" whether or not it is in equilibrium. At a single-molecule level, therefore, the distinction between equilibrium and nonequilibrium situations becomes blurred. This observation is particularly pertinent to molecular motors discussed further in this chapter. The directional motion produced by a molecular motor (e.g., the swimming of bacteria or cargo transport across the cell) is a consequence of nonequilibrium chemical composition of its surroundings. The direction of motor motion can, in principle, be reversed by changing this composition. Moreover, the trajectory followed by a motor, say, when it performs a forward step is (statistically) a time reversal of the backward step [2]. In this regard, each backward or forward step is "reversible" and so no energy has to be wasted through dissipation as the motor converts chemical energy into work. The efficiency of molecular machines will be further discussed at the end of this chapter.

The subject of single-molecule biology is too broad to address in any reasonably systematic manner in a single chapter. Instead of attempting such a hopeless task, I will try to outline how the basic physical ideas developed earlier in this book play out when applied to biomolecular transformations that sustain life and how single-molecule approaches can enhance our understanding of the living matter. Two specific examples are chosen to illustrate those ideas: the kinetics of enzyme catalysis and the motion of molecular motors. Rather than describing the molecular details in each case (which the reader can find in standard Biochemistry texts), very simple yet predictive models will be introduced. Those models can be (and have been) used to rationalize single-molecule studies of enzyme catalysis and biomolecular machines.

10.1   SINGLE-MOLECULE VIEW OF ENZYME CATALYSIS

Placed on a shelf of a grocery store, sucrose (table sugar) lasts a long time. Same sugar, when consumed during an athletic event, can be almost immediately used up to convert its chemical energy into useful (from the athlete's perspective) work. Same chemical energy can also be released by, e.g., burning the sugar, a process commonly encountered in making campfire marshmallows but also sometimes leading to sugar plant explosions.[1] Raising the temperature speeds up the oxidation of sucrose. However, since living organisms are not equipped with a machinery to achieve a temperature of over a hundred degrees of Celsius and to literally burn their fuel, they must speed up the necessary chemical transformations in a different way. The solution is provided by Nature's catalysts called enzymes.

Enzymes are (usually) proteins that reversibly bind to their targets and speed up chemical reactions without changing their reactants or products. Consider, for example, the reaction

$$R \underset{k_{P \to R}}{\overset{k_{R \to P}}{\rightleftharpoons}} P. \tag{10.1}$$

An enzyme binds the reactant R[2] and accelerates its conversion into the product, P, often by many orders of magnitude [3]. A simple yet incredibly successful model of enzyme action is provided by the Michaelis-Menten (MM) mechanism,

$$E + R \underset{k_{-1}}{\overset{k_1}{\rightleftharpoons}} ER \underset{k_{-2}}{\overset{k_2}{\rightleftharpoons}} E + P, \tag{10.2}$$

in which the binding of an enzyme to the reactant results in an intermediate complex ER. Before discussing the implications of this mechanism, we need to clarify the meaning of the rate coefficients $k_{\pm 1}, k_{\pm 2}$ appearing in Eq. 10.2. Rate coefficients were introduced in Chapter 3 in order to describe the kinetics of unimolecular reactions (i.e., reactions involving just one reactant) such as Eq. 10.1. When N_R reactant molecules are mixed with N_P product molecules (where $N_R, N_P \gg 1$), then the number of reactant molecules that are converted into the product per unit time is given by $k_{R \to P} N_R$, and, likewise, the rate of the reverse process is $k_{P \to R} N_P$, leading to the kinetic equations of the form:

$$dN_R/dt = -dN_P/dt = -k_{R \to P} N_R + k_{P \to R} N_P. \tag{10.3}$$

In practice, it is more convenient to measure the quantities of various chemicals participating in a reaction using concentrations rather than the numbers of molecules.

---

[1] In fact, the energy value or caloric content reported on food labels is commonly measured as the heat released when the food is combusted.

[2] In the biochemistry literature, the reactant is commonly called the "substrate" and is denoted as S.

The concentration of, say, the reactant, denoted [R], is simply the number of reactant molecules divided by the volume they occupy:[3]

$$[R] = N_R/V.$$

The kinetic equations describing the time evolution of concentrations are obtained by dividing Eq. 10.3 by the volume:

$$d[R]/dt = -d[P]/dt = -k_{R\to P}[R] + k_{P\to R}[P].$$

Unlike the unimolecular reaction of Eq. 10.1, a *bimolecular* reaction requires mutual encounter of *two* reactant molecules. The probability of such an event is proportional to the amounts of each reactant and thus to the product of those amounts. In particular, the rate at which the complex ER is formed in the MM scheme (Eq. 10.2) is proportional to the product [E][R]. The proportionality coefficient is the rate coefficient k_1. We note that its units are not s^{-1}, as in the unimolecular case, but $m^3 s^{-1}$. In contrast, the rate of the reverse process, where ER dissociates into the enzyme and the reactant, depends only on the concentration of ER and is given by $k_{-1}[ER]$. Same rules apply to each elementary step in the MM scheme (indicated by arrows), resulting in a set of kinetic equations of the form:

$$d[R]/dt = -k_1[E][R] + k_{-1}[ER] \tag{10.4}$$

$$d[ER]/dt = k_1[E][R] + k_{-2}[E][P] - k_2[ER] - k_{-1}[ER] \tag{10.5}$$

$$d[P]/dt = k_2[ER] - k_{-2}[E][P] \tag{10.6}$$

$$d[E]/dt = -k_1[E][R] - k_{-2}[E][P] + k_2[ER] + k_{-1}[ER]. \tag{10.7}$$

Here k_1 and k_{-2} are bimolecular reaction rate coefficients measured in $m^3 s^{-1}$ (or other appropriate units depending on the units used for concentrations) while k_{-1} and k_2 are unimolecular reaction rate coefficients measured in s^{-1}. Using similar notation for the rate coefficients that are measured in different units is usually not a problem, as the units of each coefficient are evident from the number of reactants participating in the associated process.

Since no enzyme molecules are consumed or created as a result of the overall process described by the MM scheme, the total concentration of the enzyme molecules, which are either free or bound to the reactant, must be conserved. Indeed, from Eqs. 10.4-10.7 we have $d[E]/dt = -d[ER]/dt$, or

$$d([E] + [ER])/dt = 0. \tag{10.8}$$

Because the catalysis of any particular chemical transformation is merely a step in a complex kinetic network, it is often difficult to study it directly in the cell. Instead, the reaction can be isolated in a test tube by mixing some initial amounts, $[E_0]$ and $[R_0]$, of the enzyme and the reactant and measuring the ensuing time dependence of the product and reactant concentrations. Importantly, the concentrations of the

---

[3] Chemists usually measure concentrations in moles per liter. To convert N_R into moles one divides it by Avogadro's constant $N_a \approx 6.022 \times 10^{23}$ (which, by definition, is the number of molecules in one mole).

intermediate complex [ER] or the free enzyme [E] are usually not directly detectable and so the validity of the MM mechanism can only be verified indirectly, through the predictions it makes regarding the time-dependent behavior of the reactant and product concentrations. One particularly important prediction that is commonly taken as the signature of the MM mechanism is the saturation effect: Unlike the case of the one-step process, Eq. 10.1, in the absence of the enzyme, the MM mechanism predicts that the rate at which the product is formed becomes independent of the reactant concentration [R] at sufficiently high values of [R]. This prediction can be justified in several ways, each based on different approximations. Two kinds of approximations will be examined here.

Approximation 1 (preequilibrium assumption). If certain initial amounts of the enzyme and the reactant are mixed at $t = 0$, the reactions described by Eq. 10.2 will proceed until chemical equilibrium is established. If the goal is to characterize the enzyme's efficiency to catalyze the formation of the product, we may not be particularly interested in the backward conversion of the product back to the reactant. A sensible way to characterize the forward process is to consider the *initial* rate at which the products are formed. At $t = 0$ there is no product and so the backward step $E + P \longrightarrow ER$ can be neglected, resulting in the kinetic scheme

$$E + R \underset{k_{-1}}{\overset{k_1}{\rightleftharpoons}} ER \overset{k_2}{\longrightarrow} E + P. \tag{10.9}$$

If the second step of this mechanism is slow (specifically, if $k_2 \ll k_{-1}$) then the complex ER will dissociate and reform many times before a single product-forming step will be encountered. Therefore, the concentration of ER can be estimated by assuming that chemical equilibrium (or preequilibrium) has been established in the first step, where the rate of formation of ER is balanced by the rate of its dissociation

$$k_1[E][R] = k_{-1}[ER],$$

so that

$$[ER] = \frac{k_1}{k_{-1}}[E][R]. \tag{10.10}$$

The "initial" rate of the product formation can then be estimated as

$$d[P]/dt = k_2[ER] = \frac{k_1 k_2}{k_{-1}}[E][R].$$

The quotation marks were used to distinguish this result from the true initial rate, which is zero, as, strictly speaking, no ER molecules yet exist at $t = 0$ and so no product can be formed. In contrast, the "initial" rate is the rate observed after pre-equilibrium is established but before any significant amount of product has been formed. Since the total number of enzyme molecules (whether free or forming the complex ER) is conserved (cf. Eq. 10.8), we have

$$[E] + [ER] = [E_0]. \tag{10.11}$$

Yet another approximation is usually made: that the initial concentration of the enzyme is much lower than that of the reactant,

$$[E_0] \ll [R_0].$$

If so, the number of reactant molecules tied up in the complex ER is negligible as compared to the total amount of the reactant and so we can set $[R] \approx [R_0]$. Now using Eqs. 10.10 and 10.11, we find

$$[ER] = \frac{[E_0][R_0]}{[R_0] + \frac{k_{-1}}{k_1}} \qquad (10.12)$$

and

$$d[P]/dt = \frac{k_2[E_0][R_0]}{[R_0] + \frac{k_{-1}}{k_1}}. \qquad (10.13)$$

Eq. 10.13 predicts that, at low reactant concentrations, the product formation rate is proportional to $[R_0][E_0]$, just as in a single-step bimolecular process. In contrast, a saturation behavior is seen at high concentrations of the reactant, where the reaction rate becomes independent of $[R_0]$. This behavior is easy to understand from Eq. 10.12, which shows that, at high values of $[R_0]$, essentially all the available enzyme molecules are bound and no free enzyme is available to process additional reactant. It should be emphasized, however, that the result stated by Eq. 10.13 is contingent upon not just one but several assumptions and approximations, which may or may not be satisfied in the real world.

Approximation 2 (steady-state assumption). As pointed out above, concentrations of molecules in the cell are often close to a steady state, as reactants are constantly supplied and products are removed. We thus may envisage a steady-state scenario, where the products are immediately removed once they are formed while the reactant molecules are supplied at the rate necessary to maintain the steady state. If so, the concentrations of the reactant and the enzyme-reactant complex are constant while the product concentration is zero. We thus can write

$$d[ER]/dt = k_1[E][R] - k_{-1}[ER] - k_2[ER] = 0,$$

or

$$[ER] = \frac{k_1}{k_{-1} + k_2}[E][R]. \qquad (10.14)$$

Proceeding as in the preequilibrium case but replacing Eq. 10.10 with Eq. 10.14, we now find that the product is formed at a rate equal to

$$\frac{k_2[E_0][R_0]}{[R_0] + \frac{k_{-1}+k_2}{k_1}}. \qquad (10.15)$$

Notably, Eq. 10.15 is of the same form as Eq. 10.13 and predicts similar saturation behavior.

The result stated in Equation 10.15 can be used to estimate the initial product formation rate even when no true steady-state exists: The so called steady-state approximation is usually applied when the reaction intermediate (the ER complex in the present case) is short lived (and, therefore, scarcely populated) such that it is converted almost immediately to the product once formed.

To summarize our findings about the bulk MM kinetics, two very different assumptions regarding the magnitudes of the rate coefficients and/or the conditions under which the experiment is performed lead to a similar prediction, where the rate at which an enzyme catalyzes a reaction can be written in the form

$$\text{rate} = \frac{k_2[E_0][R_0]}{[R_0] + K}. \tag{10.16}$$

As the value of the parameter K depends on the approximations made, our result is somewhat unsatisfying: Given that the mathematical form of the final answer, Eq. 10.16, is the same for two different sets of approximations, could it be more general? What if none of the above assumptions are satisfied and the kineticist is faced with solving the Equations 10.4-10.7 numerically: As these are nonlinear differential equations whose solution is quite complex, can one even define a single quantity that characterizes the product formation rate? A further concern should also be raised regarding experimental validation of the MM mechanism: If the predicted reactant concentration dependence of the rate, Eq. 10.16, is insensitive to the assumptions made regarding the underlying mechanism, does the observation of such a dependence really indicate that the MM mechanism is correct? Perhaps a variety of different mechanisms can be consistent with Eq. 10.16.

Observation of enzymatic catalysis at a single-molecule level can provide answers to the above questions. To see how, imagine monitoring the state of an individual enzyme as it cycles between the free conformation E and the complex ER. Let us not worry, for the time being, whether or not this would be a realistic experiment. When in the free state, the probability to bind a reactant during a short time interval dt is given by

$$k_1[R]dt.$$

Indeed, when multiplied by the concentration of the free enzyme, this—in accord with Eqs. 10.4 and 10.5 – will give the total amount of ER produced from the reactant R during the time dt. We can then think of the quantity

$$k_1' = k_1[R]$$

as an effective unimolecular rate coefficient describing the transition from E to ER through the binding of the reactant. Likewise, the probability for the enzyme to bind the product can be described by the effective unimolecular rate coefficient $k_{-2}' = k_{-2}[P]$. Therefore, the time evolution of a single enzyme can be represented by the following scheme:

$$E \underset{k_{-1}}{\overset{k_1'}{\rightleftharpoons}} ER \underset{k_{-2}'}{\overset{k_2}{\rightleftharpoons}} E \underset{k_{-1}}{\overset{k_1'}{\rightleftharpoons}} ER \underset{k_{-2}'}{\overset{k_2}{\rightleftharpoons}} E \rightleftharpoons \cdots. \tag{10.17}$$

Taken at face value, Eq. 10.17 has some unphysical features. In particular, it appears to imply that the conversion from ER to E can occur with two different probabilities via two distinct processes, the left-to-right transition $ER \overset{k_2}{\rightarrow} E$ and the right-to-left one, $E \overset{k_{-1}}{\leftarrow} ER$. Privy to the derivation, the reader, of course, should see no paradox here as the two transitions are not the same: The former is accompanied by the release of a product molecule while the latter results in the release of a reactant one. This fact, however, has an important consequence: Generally speaking, the two transitions should lead to different conformations of the free enzyme E. And, since the transition

E $\xleftarrow{k_{-1}}$ ER is the time-reversed transition E $\xrightarrow{k'_1}$ ER, the conformation of the enzyme at the instant the product is released is generally different from that at the start of the the turnover cycle just before the enzyme binds the reactant. Upon product release, the conformationally modified enzyme may relax back into its "typical" state before it becomes engaged in the next turnover event. There is, however, a different possibility where the new conformation is retained, resulting in a chain of events that can be described by the scheme:

$$E_1 \rightleftharpoons ER \rightleftharpoons E_2 \rightleftharpoons ER \rightleftharpoons E_3 \rightleftharpoons ER \dots. \tag{10.18}$$

The result is that, under appropriate conditions, our enzyme will undergo directional motion in the conformational space acting as a "motor." This idea will be further explored in the next section.

A further (seemingly) unphysical feature of Eq. 10.17 is that it violates the detailed balance conditions formulated in Chapter 4 (cf. Eqs.4.23 and 4.26). Indeed, detailed balance would, for example, require the ratio k'_1/k_{-1} to be determined by thermodynamic properties of the molecules E and ER alone. But an experimenter can vary k'_1 at will by changing the reactant concentration.[4] Eq. 10.17 (apparently) contradicts detailed balance only because it does not keep track of the state of the enzyme's environment, particularly of the amounts of the currently present reactant and product. If, for example, the enzyme is placed in a solution containing large quantities of reactant R and no product, it will proceed to catalyze the conversion of the reactant molecules to the product. In terms of the scheme of Eq. 10.17, this means overall motion from left to right. As more product molecules are created as a result of the enzyme action, the reverse process will kick in and the enzyme will be occasionally seen to catalyze the conversion of product molecules back to the reactant. Eventually, chemical equilibrium will be attained, where, on the average, equal numbers of reactant-to-product and product-to-reactant transitions will be taking place, precisely as detailed balance, when applied to the scheme Eq. 10.17, would require. The initially observed unidirectional motion is, therefore, driven by the nonequilibrium composition of the enzyme's environment as it evolves toward chemical equilibrium. These observations clarify the usage of master equations such as Eq. 4.25, which apparently violate both detailed balance and time reversal symmetry. They only appear to do so because they leave part of the story out of the picture.

At the single-molecule level, the enzyme's aptitude to catalyze the conversion to the products can be measured as the number of product molecules generated per unit time. X. Sunney Xie's group, for example, tethered a single enzyme to the surface of a bead and placed it within the focal spot of a laser [4]. Whenever the fluorescent product molecule was released from the enzyme, a burst of photons was observed. The number of such bursts per unit time thus yields the rate at which the enzyme catalyzes the product. Since the activity of a single enzyme can hardly influence the bulk concentrations of the reactant and product in solution, it is reasonable to assume the effective rate coefficients k'_1 and k'_{-2} to remain constant for the duration of the

---

[4] The ratio k_1/k_{-1}, in contrast, *is* determined by the thermodynamic properties of the molecules R, E, and ER.

experiment. Moreover, the experiment was performed under the conditions where the product concentration is negligible so we can set $k'_{-2} \approx 0$. With these assumptions, the enzyme's kinetics can be approximated by the scheme

$$E \underset{k_{-1}}{\overset{k'_1}{\rightleftharpoons}} ER \overset{k_2}{\longrightarrow} E \underset{k_{-1}}{\overset{k'_1}{\rightleftharpoons}} ER \overset{k_2}{\longrightarrow} E \rightleftharpoons \cdots . \tag{10.19}$$

Let w_{ER} be the probability to find the enzyme bound to the reactant and $w_E = 1 - w_{ER}$ the probability that it is free. These probabilities should not change with time and so, according to the scheme of 10.19, we have

$$dw_{ER}/dt = -dw_E/dt = k'_1 w_E - k_{-1} w_{ER} - k_2 w_{ER} = 0$$

and

$$w_{ER} = \frac{k'_1 w_E}{k_2 + k_{-1}}.$$

Combining this with the requirement that the two probabilities must add up to one, $w_E + w_{ER} = 1$, we find

$$w_E = \frac{1}{1 + \frac{k'_1}{k_{-1}+k_2}}, \quad w_{ER} = \frac{1}{1 + \frac{k_{-1}+k_2}{k'_1}}.$$

The rate at which product molecules are released (which we will call v) is simply k_2 multiplied by the probability of forming the complex ER:

$$v = k_2 w_{ER} = \frac{k_2}{1 + \frac{k_{-1}+k_2}{k'_1}} = \frac{k_2[R]}{[R] + K}, \tag{10.20}$$

where

$$K = \frac{k_2 + k_{-1}}{k_1}. \tag{10.21}$$

Equation 10.20 is the single-molecule analog of the Michaelis-Menten result, Eq. 10.16. It differs from the quasiequilibrium approximation (Eq. 10.13) but is analogous to the steady-state result (Eq. 10.15), which is not surprising given the experimental conditions that ensure that (single) enzyme action does not affect the reactant and product concentrations.

Since the turnover rate v can be calculated as the number of enzymatic turnover events n (or the number of experimentally observed photon bursts) normalized by the measurement time τ, its inverse,

$$\langle t \rangle = \tau/n,$$

is the the mean turnover time (i.e., the average time between two product release events). But a single-molecule experiment is not limited to measuring the mean value of the time t. Rather, it can give us its entire probability distribution, $w(t)$. Could we learn more about how the enzyme works if we know $w(t)$? The answer is a resounding yes. It turns out that, while certain essential details of enzymatic action are fundamentally impossible to infer from the mean turnover time, the same details are readily inferred (at least in principle) from the time distribution.

To illustrate this point, imagine that the enzyme-catalyzed conversion of the reactant to the product involves not just a single intermediate ER, as in the MM scheme, but multiple intermediate states. We then replace Eq. 10.17 by a multistep scheme of the form

$$E \xrightarrow{k_1'} I_1 \xrightarrow{k_2} I_2 \rightarrow \cdots \xrightarrow{k_N} I_N \xrightarrow{k_{N+1}} E \xrightarrow{k_1'} I_1 \ldots. \qquad (10.22)$$

To simplify calculations, all the steps in the scheme are assumed to be irreversible. It is further assumed that the first step represents the enzyme binding to the reactant, with an effective rate coefficient of $k_1' = k_1[R]$. This step is followed by a series of conformational rearrangements of the ER complex, via a series of intermediates I_1, I_2, ..., and, ultimately, by the release of the product occurring in the step $I_N \rightarrow E$. Let t_1 be the dwell time in the state E before the transition to I_1 occurs, t_2 the time between the moment the molecule arrives in I_1 and the moment it makes a transition from I_1 to I_2, and so forth. Finally t_{N+1} is the time between the system's arrival in I_N and the release of the product accompanied by the transition back to the free form E. The total turnover time is

$$t = t_1 + t_2 + \cdots + t_{N+1}$$

and its average is

$$\langle t \rangle = \langle t_1 \rangle + \langle t_2 \rangle + \cdots + \langle t_{N+1} \rangle.$$

Since the only escape pathway from any state is a transition to the next state, the mean dwell time is simply the inverse of the escape rate coefficient (Chapter 3). Thus we find $\langle t_1 \rangle = 1/k_1'$, $\langle t_i \rangle = 1/k_i$ for $i > 1$, and

$$\langle t \rangle = \frac{1}{k_1'} + \sum_{i=2}^{N+1} \frac{1}{k_i}.$$

Taking the inverse, we find that the average turnover rate v still obeys a Michaelis-Menten-type equation,

$$v = \langle t \rangle^{-1} = \frac{[R]\tau_2^{-1}}{[R] + K}, \qquad (10.23)$$

provided that

$$\tau_2 = \sum_{i=2}^{N+1} \frac{1}{k_i}$$

is the mean time between the arrival in I_1 and the product release, while the Michaelis-Menten constant K is now redefined as

$$K = \tau_2^{-1}/k_1.$$

I will leave it to the reader to verify that the alternative derivation of the rate v as the product $k_{N+1}w_N$, where w_N is the steady-state probability of finding the enzyme in the state I_N, gives the same result.

The upshot of the above analysis is that the validity of the MM equation tells the experimenter nothing about the actual number of intermediates encountered in the enzymatic process. Things are, however, even worse. The actual mechanism may involve reversible steps, pathways occurring in parallel or even convoluted networks of states, yet the MM equation would still remain correct, provided that its parameters are appropriately defined [5,6]! This suggests that the reason for the overwhelming success of the MM equation has nothing to do with the validity of the originally assumed Michaelis-Menten scheme but is rather due to the insensitivity of the [R]-dependence of the turnover rate to the specifics of the mechanism of enzyme action.[5]

In contrast, alternative mechanisms can be differentiated by examining the distribution $w(t)$. In particular, the number N of intermediates in the scheme 10.22 affects the short-time behavior of $w(t)$, which is given by a power law,

$$w(t) \propto t^N \tag{10.24}$$

in the limit $t \to 0$. For example, in a single-step process ($N = 0$), such as the direct reaction described by Eq. 10.1, we have (cf. Chapter 3) $w(t) = k_{R \to P} e^{-k_{R \to P} t}$, and so the probability distribution approaches a constant value $w(t) \propto t^0$ at $t \to 0$. A single intermediate, on the other hand, results in a $w(t)$ that vanishes at short times, when the intermediate is not yet populated. We can find $w(t)$ in this case by noticing that its calculation is completely analogous to the derivation of Eq. 7.14 in Chapter 7. Indeed, Eq. 7.14 described the probability distribution of the time taken by a two-step process (the first step being excitation of the molecule and second step the emission of a photon), where the two steps were statistically independent and the time of each step had an exponential distribution. Replacing the rate coefficient for the first step, k_{ex}, by k_1' and the rate coefficient for the second step, Γ, by k_2, we obtain

$$w(t) = \frac{e^{-k_1' t} - e^{-k_2 t}}{(k_1')^{-1} - k_2^{-1}}.$$

This function vanishes in the limit of the vanishing time t. Moreover, expanding this in a Taylor series to lowest order in t we find

$$w(t) \approx k_1' k_2 t,$$

in accord with Eq. 10.24. Since the MM scheme is the case of a single, intermediate observation of a linear dependence $w(t)$ at short times would lend support to the MM mechanism.

The derivation of Eq. 7.14 can be extended to find the short-time limit of the probability distribution for an irreversible multi-step process of Eq. 10.22, which, indeed, agrees with Eq. 10.24. It turns out, however, that a much more general claim can be made [7]: Suppose the enzyme action is described by some complex kinetic network, such as the one shown in Figure 10.1. Starting in the free enzyme state E, we measure the time t it takes to reach the state E′ where the product is released. Then

---

[5] At the same time, deviations from MM kinetics can, in principle, provide information about this mechanism.

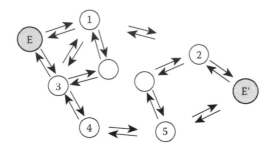

FIGURE 10.1 If enzymatic turnover proceeds via a complex kinetic network involving many steps, some network properties (such as the length of the shortest path connecting the initial (E) and the final (E') states) may be inferred from the distribution of the first passage time t between these states. Here, the shortest path involves 3 steps and, consequently, this distribution is proportional to t^2 at $t \to 0$.

the short-time behavior of its probability distribution, $w(t)$, is given by Eq. 10.24, where $N+1$ is the *smallest number of elementary kinetic steps* connecting E and E'. This result holds regardless of whether or not the network contains irreversible steps.

In Fig.10.1, for example, the shortest pathway is going through the intermediates 1 and 2 in three steps and so we expect $w(t) \propto t^2$. This conclusion appears counterintuitive, as it is independent of the probabilities of each step. Imagine, for example, that each step in the 4-step pathway proceeding through intermediates 3, 4, and 5 has rate coefficients that are much higher than those for the 3-step pathway. Still, the shortest pathway dominates if the time t is short enough, resulting in a t^2 dependence.

This result can be proven as follows. The probabilities to occupy each state in the network obey the master equation of the form of Eq. 4.20. Let us consider a specific pathway that starts in E at zero time, ends in E' at a time t, and proceeds through intermediates $i_1, i_2, \ldots, i_n$. Furthermore, suppose that the transition to state i_1 has occurred between t_1 and $t_1 + dt_1$, the transition to i_2 between t_2 and $t_2 + dt_2$, and so forth. What is the probability of such a chain of events? The answer is a simple generalization of the arguments made in Chapters 3 and 7. The probability that the first transition occurred between t_1 and $t_1 + dt_1$ is the probability $w_S(t_1)$ to survive in the initial state (E) up to the time t_1 multiplied by the transition probability $k_{E \to i_1} dt_1$. For the purpose of calculating the survival probability in E, the escape process from E can be described by the one-step kinetic scheme,

$$E \xrightarrow{k_{E \to}} \text{adjacent states}$$

where

$$k_{E \to} = \sum_j k_{E \to j}$$

is the total escape rate (coefficient) from the state E. This gives

$$w_S(t_1) = e^{-k_{E \to} t_1}.$$

We can now multiply the result by the probability of the second transition occurring between t_2 and $t_2 + dt_2$ and so forth. The result is:

$$k_{E \to i_1} dt_1 e^{-k_{E \to} t_1} k_{i_1 \to i_2} dt_2 e^{-k_{i_1 \to}(t_2 - t_1)} \dots k_{i_n \to E'} dt_n e^{-k_{i_n \to}(t - t_n)}. \tag{10.25}$$

In the limit $t \to 0$ (and, correspondingly, $t_i \to 0, i = 1, \dots, n$) we can replace all the exponentials by 1 so that the above probability becomes

$$k_{E \to i_1} dt_1 k_{i_1 \to i_2} dt_2 \dots k_{i_n \to E'} dt_n. \tag{10.26}$$

To calculate the probability of arriving in the state E′ at time t we must integrate over all intermediate times subject to the constraint $0 < t_1 < t_2 < \cdots t_n < t$ and then sum over all possible sequences of states that lead to the final state E′. Each sequence involving n intermediates will then contribute a term proportional to

$$\int_{0 < t_1 < t_2 < \cdots < t_n < t} dt_1 dt_2 \dots dt_n = t^n / n!$$

The lowest order (and, therefore, dominant) term in the Taylor expansion of $w(t)$ will thus come from the shortest path(s) connecting E and E′, which proves our claim.

In practice, accurate determination of the short-time behavior of $w(t)$ could be challenging precisely because, as predicted by Eq. 10.24, the probability to observe fast turnover events vanishes quickly as t is decreased. The above exercise should nevertheless convince us that certain features of the network topology are imprinted in the properties of the distribution $w(t)$ and that a clever experimenter can infer at least some details of the underlying mechanism by analyzing this distribution.

10.2   ENZYMES AS MOLECULAR MOTORS

Earlier in this chapter, the possibility was suggested that a single enzymatic turnover event, as described by the Michaelis-Menten schemes of Eqs.10.17 and 10.18, may result in a conformational change of the enzyme itself. If this change amounts to a spatial displacement of the enzyme in some preferred direction, repeated turnover events will lead to directed motion. Although no preferred direction can possibly emerge for an enzyme that is oriented randomly in space, directionality of motion can be ensured when the enzyme is attached to a one-dimensional "track" (such as a microtubule) in the cell. The enzyme then becomes a molecular motor.

To investigate the motor motion, we envisage a simple model, where a forward step of certain length d occurs every time a reactant molecule R is converted to the product P (see Figure 10.2, where left to right is chosen to be the forward direction). The reactant can thus be regarded as a "fuel" molecule, whose consumption results in forward motion. In the presence of the product, the reverse process is also possible, where a product molecule P binds to the enzyme and is converted into R: Being the time-reversal of a forward step, this event must then be accompanied by a displacement of the same length d to the left. Unlike artificial engines that maintain steadfast progress, biological motors thus tend to move in an unsteady fashion, with an occasional back step interrupting the forward motion. Nonequilibrium composition of the motor's

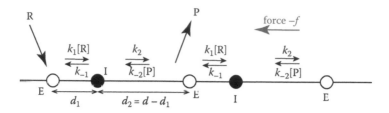

FIGURE 10.2 A simple model of molecular motor consists of an enzyme that catalyzes the conversion of the "fuel" reactant molecule to the product. In a single forward step one molecule R is converted to P and the enzyme makes a step of a total length of d to the right along its track. The reverse process where the enzyme binds P, which is then converted to R, results in a step of the same length to the left.

environment is clearly essential for its operation. Indeed, if equilibrium concentrations [R] and [P] were present in the cell, the motor would, on the average, go nowhere as the average number of steps in either direction would be the same.

Although, for the purpose of the present discussion, the precise nature of the molecules R and P is unimportant, it should be pointed out that the specific fuel molecule that powers the majority of molecular motors is adenosine triphosphate, or ATP. The enzyme, or motor, catalyzes the hydrolysis of ATP to produce adenosine diphosphate (ADP) and an inorganic phosphate, which can be regarded as waste products of the motor action. The simple motor model depicted in Fig.10.2 is imprecise in that it does not show other molecules (water, inorganic phosphate) involved in the process. Nevertheless, it captures the essential physics and will be used here without further attempts to make it more realistic.

Left to its own devices, the motor will walk (on the average) forward or backward, without performing any useful work. The mean velocity of its motion can be estimated in a manner similar to the calculation of the mean enzymatic turnover rate from the preceding section, except that the possibility of a backward step must now be taken into account. The intermediate state I (which is the same as the ER state in the previous section) divides each motor step into two substeps (Fig.10.2). Let us assume that the first substep is of length d_1 and, correspondingly, the second substep as a length of $d_2 = d - d_1$. From each of its states, E or I, the motor can step either right or left. The average frequencies of right and left steps occurring from E are, respectively, $k_1' w_E$ and $k_{-2} w_E$, where w_E, as before, is the steady-state probability to be in the free state E. Similarly, right and left steps from the intermediate state occur, on the average, with the frequencies equal, respectively, to $k_2 w_I$ and $k_{-1} w_I$, where w_I is the steady-state probability of being in I. The average motor displacement during a time Δt is thus given by

$$\Delta x = \Delta t [k_1' w_E d_1 - k_{-2}' w_E d_2 + k_2 w_I d_2 - k_{-1} w_I d_1]$$

and the average velocity is

$$v = \Delta x / \Delta t = k_1' w_E d_1 - k_{-2}' w_E d_2 + k_2 w_I d_2 - k_{-1} w_I d_1. \qquad (10.27)$$

The probabilities w_E and w_I can be found from the condition that they do not change in time,

$$dw_E/dt = -dw_I/dt = k_2 w_I - k'_{-2} w_E - k'_1 w_E + k_{-1} w_I = 0,$$

and from the conservation of probability,

$$w_E + w_I = 1,$$

resulting in

$$w_E = \frac{k_2 + k_{-1}}{k'_1 + k'_{-2} + k_2 + k_{-1}}$$

and

$$w_I = \frac{k'_1 + k'_{-2}}{k'_1 + k'_{-2} + k_2 + k_{-1}}.$$

Substituting these results into Eq. 10.27, we obtain

$$v = d \frac{k'_1 k_2 - k_{-1} k'_{-2}}{k'_1 + k'_{-2} + k_2 + k_{-1}}. \tag{10.28}$$

The velocity becomes equal to zero if

$$k'_1 k_2 = k_{-1} k'_{-2}. \tag{10.29}$$

This condition corresponds to chemical equilibrium among all the chemical species involved in the process (i.e., R, P, E, I). Indeed, detailed balance (see Chapter 4) requires that, in equilibrium, the rate of every elementary reaction should be equal to the rate of its reverse. Then we have

$$k'_1 w_E = k_{-1} w_I$$

and

$$k_2 w_I = k'_{-2} w_E$$

If we multiply these two equations together, we get

$$k'_1 k_2 w_E w_I = k_{-1} k'_{-2} w_E w_I,$$

which is equivalent to Eq. 10.29. To characterize the degree to which our system deviates from equilibrium, it is helpful to consider the ratio

$$\chi = \frac{k'_1 k_2}{k_{-1} k'_{-2}},$$

or, recalling the definition of the effective rate coefficients k'_1 and k'_{-2},

$$\chi = \frac{k_1 [R] k_2}{k_{-1} k_{-2} [P]} = \frac{[R]\,[P]_{eq}}{[P]\,[R]_{eq}}, \tag{10.30}$$

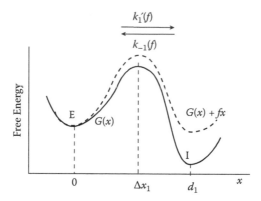

FIGURE 10.3 Similarly to Chapter 8, the effect of the force on the molecular motor can be understood using a simple one-dimensional model, where a force $-f$ acting toward the left tilts the free energy landscape, thereby decreasing the rate coefficient k_1' for the forward step and increasing the rate coefficient k_{-1} for the backward step.

where $[R]_{eq}$, $[P]_{eq}$ are the equilibrium concentrations of R and P corresponding to $\chi = 1$. When the parameter χ is much greater than 1, backward steps are much less likely than the forward ones, resulting in nearly unidirectional motion. Likewise, if $\chi \ll 1$, the motor is walking predominantly backwards. When χ is close to 1, the motor surroundings are near chemical equilibrium, providing only a small bias to move in a certain direction.

Exercise

Consider even a simpler (albeit unphysical) model of a molecular motor, which jumps between equivalent sites denoted E and separated by a distance d according to the scheme:

$$\mathrm{E} \underset{k_{-1}}{\overset{k_1}{\rightleftharpoons}} \mathrm{E} \underset{k_{-1}}{\overset{k_1}{\rightleftharpoons}} \mathrm{E} \underset{k_{-1}}{\overset{k_1}{\rightleftharpoons}} \cdots .$$

What is the average velocity of this motor?

The motion of the motor considered so far did not result in any useful work. Suppose now that some cargo is attached to the motor. To move it forward (i.e., from left to right in our scheme), the motor must exert, on the average, some force f. Equivalently, the cargo exerts a force $-f$ on the motor that opposes the motor's forward motion (Fig. 10.2). Such a force will help the motor jump to the left and oppose its transition to the right, thereby reducing its average velocity. The same scenario can also be reproduced *in vitro*, where a desired force can be exerted on the motor using the single-molecule pulling techniques described in Chapters 2 and 8. In addition to the force dependence of the average motor velocity, $v(f)$, single-molecule measurements can provide other information. For example, similarly to the case of single-molecule enzyme studies, the statistics of the times at which the motor steps in either direction can elucidate details of its action. The following discussion will, however, be limited to the dependence $v(f)$, as this dependence quantifies the strength, power, and efficiency of the molecular machine. A discussion of other measurable properties of motors and their comparison with theoretical predictions can be found, e.g., in ref. [9] and the references therein.

The curve $v(f)$ can be calculated from Eq. 10.28 if the force dependence of each of the rate coefficients is known. The problem of estimating the force effect on the rate of a chemical process was discussed in Chapter 8. Consider, for instance, the first substep performed by the motor, where conversion of the free enzyme E into I results in a step of length d_1 to the right (Fig.10.2). The states E and I correspond to two minima of the free energy $G(x)$ viewed as a function of the motor position x along its track (Figure 10.3). The force tilts the curve $G(x)$ raising the barrier to the direct reaction (from left to right) and lowering the barrier of the left one. If the force-induced shifts in the positions of the minima and the barrier are negligible, then the force effect on the rate coefficients can be approximated by the Eyring-Zhurkov-Bell formula (Eq. 8.24):

$$k_1'(f) = k_1'(0)e^{\frac{-f\Delta x_1}{k_B T}}$$

and, similarly,

$$k_{-1}(f) = k_{-1}(0)e^{\frac{f(d_1 - \Delta x_1)}{k_B T}}.$$

Here Δx_1 denotes the position of the barrier relative to that of the left minimum corresponding to E, as shown in Fig. 10.3. Thus the force raises the barrier to the forward process by $f\Delta x_1$ and lowers the barrier to the reverse process by $f(d_1 - \Delta x_1)$. Analogous equations can be written for the rate coefficients in the second step that is of length d_2:

$$k_2(f) = k_2(0)e^{\frac{-f\Delta x_2}{k_B T}}$$

and

$$k_{-2}'(f) = k_{-2}'(0)e^{\frac{f(d_2 - \Delta x_2)}{k_B T}},$$

where the transition-state displacement Δx_2 is defined similarly. Substitution of these rate coefficients into Eq. 10.28 now gives the force dependence of the motor speed, $v(f)$. Let us write the result in the following form:

$$v(f) = d\omega(f)[\chi(f) - 1], \qquad (10.31)$$

where

$$\omega(f) = \frac{k_1(f)'k_2(f)}{k_1'(f) + k_{-2}'(f) + k_2(f) + k_{-1}(f)}$$

and

$$\chi(f) = \frac{k_1'(f)k_2(f)}{k_{-1}(f)k_{-2}'(f)} = \chi(0)e^{-\frac{fd}{k_B T}}. \qquad (10.32)$$

The precise shape of the force-vs.-velocity curve $v(f)$ predicted by our theory depends on the choice of eight parameters (i.e., zero-force rate coefficients and displacements). A systematic exploration of this dependence can be found in the literature [8]. Some of the salient features of this dependence can, however, be understood without doing any numerical calculations. Specifically, the velocity $v(f)$ is expected to decrease

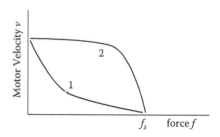

FIGURE 10.4 Application of an opposing force lowers the average velocity of a molecular motor until it stalls at a certain value of the force f_s. Different curves $v(f)$ are possible, depending on the kinetic details of the motor action. Curve 1 exemplifies a weak and inefficient motor while Curve 2 corresponds to a strong(er) and (more) efficient motor.

as the force is increased until the motor stalls at some value of the force, f_s, where $v(f_s) = 0$ (Figure 10.4). At the stall force f_s, the motor is equally likely to step forward and back. A force that exceeds f_s pushes the motor in the backward direction. The mechanical work exerted by the force on the motor in this case drives the synthesis of the fuel molecules R from P.

In evaluating macroscopic machines, we usually want to know how strong, powerful, or efficient they are. While interdependent, these criteria are not equivalent. For example, the engine torque characterizes a car's ability to pull heavy loads at low speeds. The engine horse power, on the other hand, provides a better measure of its ability to pass another car on a freeway. The car engine efficiency (measured, e.g., in miles per gallon) further depends on the manner in which the car is driven and, in particular, on its speed. The importance of each of these parameters is determined by the driver's priorities. Similar considerations apply to molecular motors, where biological function is likely to dictate constraints on their power, strength and efficiency. For example, high force needs to be produced by a motor that packs a virus DNA into its protein shell, as this force must overcome electrostatic and other interactions encountered in this process. In contrast, a cargo transporting motor may not necessarily need to be strong, as the resistance force exerted by a typical cargo is relatively low.

To quantify these properties, the energetic value of the "fuel" molecules used by the motor needs to be considered. At first glance, the free energy difference,

$$\Delta G = G_P - G_R,$$

between the product and the reactant molecules appears to be an adequate measure for this value. In the spirit of Chapter 9, one expect the free energy difference $-\Delta G$ to be available to the motor for performing mechanical work (provided that $\Delta G < 0$). A closer inspection of Eq. 10.31 along with Eqs.10.30 and 10.32 suggests that the value of ΔG cannot be telling the whole story. Indeed, using Eq. 10.30, we have

$$\chi(0) = \frac{[R]}{[P]}\frac{q_P}{q_R} = e^{-\frac{\Delta G}{k_B T}}\frac{[R]}{[P]}.$$

Here, we have used the fact that equilibrium concentration must agree with Maxwell-Boltzmann statistics and so the ratio $\frac{[P]_{eq}}{[R]_{eq}}$ must be equal to the ratio of the product (q_P) and reactant (q_R) partition functions (see Chapters 4–5 and Appendix B), which, in turn, is related to their free energy difference through the identity

$$\Delta G = -k_B T \ln \frac{q_P}{q_R}.$$

Now using Eq. 10.32, we find

$$\chi(f) = e^{-\frac{\Delta G + fd}{k_B T}} \frac{[R]}{[P]}. \tag{10.33}$$

This equation indicates that if the conversion R $\longrightarrow$ P is thermodynamically unfavorable (i.e., $\Delta G > 0$) the motor can still move forward and perform mechanical work provided that the concentration of the product is low enough as compared to that of the reactant. In other words, our motor could, in principle, pull on its cargo and, simultaneously, synthesize a high-energy product P from a low-energy molecule R![6] Likewise, Eq. 10.33 predicts the motor to adopt a certain preferred direction of motion even when $\Delta G = 0$, provided that the reactant and the product concentrations are different.

A more careful analysis accounts for the apparent free energy deficit. Consider, for example, the case $\Delta G = 0$. The equilibrium reactant and product concentrations are then equal. If the initial concentrations are not, they will spontaneously evolve (through the interconversion between R and P) until equal concentrations are reached. From a thermodynamic standpoint, this process is driven by the increase in the *entropy* of the R/P mixture (and, consequently the decrease in its free energy). To account for this mixing entropy, consider the partition function $q(N_R)$ of a solution that consists of N_R reactant molecules and $N_P = N - N_R$ products molecules, N being the constant total number of molecules. Each molecule can be either in the reactant or the product state. If we assign N_R specific molecules to be in the reactant state and N_P specific molecules to be in the product state, the resulting partition function is the product of individual partition functions (see Appendix B), which is equal to $q_R^{N_R} q_P^{N_P}$. But since, rather than the specific state of each molecule, all we care about is the total number of molecules in each state, we need to account for the total number of ways N_R molecules can be chosen from the total of N molecules. If we think of molecules as being placed in two boxes, one labeled "R" and the other "P," we need to know the total number of ways we can distribute N molecules between the two boxes such that N_R molecules are in the first box and the rest of them in the second. The answer is $\frac{N!}{N_R!(N-N_R)!}$ ways, where the exclamation sign denotes the factorial (e.g., $N! = 1 \times 2 \times \ldots \times N$).[7] The

---

[6] A high positive value of ΔG would, however, also imply a high free energy barrier to the formation of the product, which would make our motor slow.

[7] This result can be obtained as follows: Pick the first molecule from the N possibilities and place it into the first box. Pick the second one out of the $N - 1$ remaining possibilities and place it into the first box (provided that $N_R > 1$). Keep on adding molecules to the first box until it contains N_R molecules. Now pick the next molecule out of the $N - N_R$ possibilities and place it in the second box. Proceed filling up the second box until the last molecule is used. The total number of ways to accomplish this is equal

partition function we seek is then

$$q(N_R) = \frac{N!}{N_R!(N - N_R)!} q_R^{N_R} q_P^{N_P}.$$

Now let us consider the change in the free energy of the solution when the number of reactant molecules is decreased by 1 (and, correspondingly, the number of product molecules is increased by 1). This can be written as

$$\Delta\mu = -k_B T \ln \frac{q(N_R - 1)}{q(N_R)} = -k_B T \ln \frac{N_R q_P}{(N_P + 1)q_R} \approx -k_B T \ln \frac{q_P[\text{R}]}{q_R[\text{P}]}. \quad (10.34)$$

Unlike ΔG, the concentration-dependent quantity $\Delta\mu$, referred to as the *chemical potential* difference between P and R, accounts for the total solution entropy and provides the measure of the maximum work, per step, available to the motor. In terms of this quantity, Eq. 10.33 can now be rewritten as

$$\chi(f) = e^{-\frac{\Delta\mu + fd}{k_B T}}. \quad (10.35)$$

We are now in position to quantify the motor strength, power, and efficiency in terms of the thermodynamic value of the motor fuel quantified by $\Delta\mu$. The strength is readily calculated from our theory as the maximum force that the motor can sustain while moving forward. In other words, it is the stall force f_s, which can be found from the condition $\chi(f_s) = 1$, yielding a linear relationship between the free energy input and the motor strength:

$$f_s = -\Delta\mu/d. \quad (10.36)$$

The motor power is the product of the exerted force and its speed,

$$P = f v(f).$$

It is zero both in the absence of an external load, $f = 0$, and at the stall force $f = f_s$, where the speed becomes zero. It is then obvious that P attains a maximum value at some intermediate value of the force.

Finally, the efficiency ϵ of the motor can be defined as the ratio of the work performed by the motor in a single step and the available free energy per fuel molecule:

$$\epsilon = \frac{fd}{|\Delta\mu|}. \quad (10.37)$$

According to Eqs. 10.37 and 10.36, the motor efficiency approaches its thermodynamic limit $\epsilon = 1$ when the force f approaches the stall force f_s. Unfortunately, the motor

---

to $N \times (N - 1) \times \ldots \times 1 = N!$. However this expression overcounts the total number of possibilities because we do not care in which order the molecules that ended up, e.g., in the first box, were placed. If we enumerate the molecules in this box as $1, 2, \ldots, N_R$ then all sequences such as $2, 1, \ldots, N_R, N_R, 1, \ldots, 2$ or any other permutation of these numbers should count as the same. How many permutations are there? The answer is, again, $N_R!$: the first number in the sequence can be picked in N_R ways, the second in $N_R - 1$ ways, and so on. To correct for the ovecounting that results from ordering the molecules within the first box, we thus divide $N!$ by $N_R!$. Likewise we divide it by $(N - N_R)!$ to correct for overcounting the various ways of placing the molecules in the second box.

also becomes infinitely slow and produces zero power in this limit. Both the efficiency and the power are zero in the opposite limit of $f = 0$, where the motor does no useful work. In most practical cases, the efficiency at some intermediate value of the force f or velocity v, corresponding to the conditions in the cell, would be of interest. It is then clear that a concave dependence $v(f)$ exemplified by curve 2 in Fig. 10.4 would correspond to a motor that is more powerful and efficient (given the same velocity) then a motor described by the convex curve 1. As shown in [8], interplay of the model parameters can lead to a variety of scenarios including both concave and convex $v(f)$, as well as more complex dependencies. Therefore even the simple two-state model of Fig.10.2 can produce rich behavior and provide clues as to how to design efficient motors. Understanding biological molecular motors as well as designing artificial ones is currently an active research field (see, e.g., [9] for a review). Likewise, many other cellular phenomena are now amenable to single-molecule studies. A single chapter in this book cannot possibly do justice to the fascinating and diverse field of single-molecule biophysics and the discussion above only attempts to scratch the very surface. However if this discussion encourages the reader to look into the subject further I will consider my job done.

REFERENCES

1. L.D. Landau and E.M. Lifshitz, *Statistical Physics*, Elsevier, 1980.
2. R. Dean Astumian, "Microscopic reversibility as the organizing principle of molecular machines", *Nature Nanotechnology*, vol. 7, p. 684, 2012.
3. Alan Fersht, *Structure and Mechanism in Protein Science*, W.H. Freeman and Company, New York, 1999.
4. Brian P. English, Wei Min, Antoine M. van Oijen, Kang Taek Lee, Guobin Luo, Hongye Sun, Binny J. Cherayil, S. C. Kou, and X. Sunney Xie, "Ever-fluctuating single enzyme molecules: Michaelis-Menten equation revisited", *Nature Chemical Biology*, vol.2, p. 87, 2006.
5. Jianshu Cao, "Michaelis-Menten equation and detailed balance in enzymatic networks", *J. Phys. Chem. B*, vol. 115, p. 5493, 2011.
6. Anatoly B. Kolomeisky, "Michaelis-Menten relations for complex enzymatic networks", *J. Chem. Phys.*, vol. 134, 155101, 2011.
7. Xin Li and Anatoly B. Kolomeisky, "Mechanisms and topology determination of complex chemical and biological network systems from first-passage theoretical approach", *J. Chem. Phys.* vol. 139, 144106, 2013.
8. Michael E. Fisher and Anatoly B. Kolomeisky, "The force exerted by a molecular motor", *Proc. Natl. Acad. Sci. USA*, vol. 96, p. 697, 1999.
9. Anatoly B. Kolomeisky, "Motor proteins and molecular motors", *J. Phys. Condens. Matter*, vol. 25, 463101, 2013.

A Probability Theory, Random Numbers, and Random Walks

A.1   RULES FOR CALCULATING PROBABILITIES

Tossing a coin is commonly used as an example of a random process, with the two outcomes, heads (H) and tails (T), being equally likely. We will use this experiment to illustrate some of the essential mathematical rules that are used to describe random processes. Suppose we toss a coin $N \gg 1$ consecutive times. We can write the outcome as a string of letters:

$$\underbrace{\text{HHHTTTHTTHTHTTH}\dots}_{N}.\tag{A.1}$$

A practical way to measure the probability of one of the outcomes, say heads, is to divide the number of H's in the string by the total string length:

$$w(\text{H}) = \frac{\text{number of H's in the string}}{N} \approx 1/2 \tag{A.2}$$

since we expect that if N is large enough, about half of the letters in the string will be H's.

We can also define the probability of generating a certain sequence of H's and T's by repeatedly tossing the coin. For example, consider the following two sequences, each containing 10 letters:

$$\text{HHHHHHHHHH}$$

$$\text{HTTHHTHTTH}$$

It may appear to that the first sequence is very "unlikely," in contrast to the more random-looking second one. This is not so. Each of these sequences is completely unique and is equally probable (since H and T are equally likely). To estimate the probability for any given sequence, we can imagine generating random 10-letter sequences, by performing 10 coin tosses, over and over again. Every once in a while the right sequence will be accidentally generated: The fraction of such instances relative to the total number of sequences is the probability that we seek. Since all sequences are equally likely, we find that the probability of each individual sequence is

$$(1/2)^{10} = \frac{\text{one particular sequence}}{2^{10} \text{ possible sequences}}.$$

It is further reasonable to assume that the outcome of a single coin toss would not be influenced in any way by our earlier coin tossing activities. That is, regardless of the previously generated sequence of H's and T's, next time we toss the coin, the two possible outcomes will still be equally likely. If this condition holds then we say that the coin tosses are statistically independent. *If two random events (such as coin tosses) are statistically independent, then the probability w(B,A) that their outcomes are A and B is the product of the individual probabilities of those outcomes*

$$w(B,A) = w(B)w(A). \tag{A.3}$$

For example, possible outcomes of a two-coin-toss experiment are HH, TH, HT, and TT. Since they are equally likely, the probability of each outcome is 1/4. We could obtain this result from the above multiplication rule, e.g.,

$$w(HH) = w(H)w(H) = (1/2) \times (1/2) = 1/4.$$

But not all events are statistically independent. Consider the following scenario. Imagine that you and I have a loaf of bread and we know that a single coin is baked into it (but do not know where). According to many cultures, finding a coin in your bread means luck. We divide the loaf into two equal parts and proceed to find out which of us is lucky. The probability that I have the coin in my half (let us call this event A) is $w(A) = 1/2$. Likewise, the probability that you are lucky to find the coin (call it B) is $w(B) = 1/2$. But the probability $w(B, A)$ that we both find the coin is not $w(B)w(A) = 1/4$ but zero! Let now B' correspond to you *not* finding the coin. The probability $w(B', A)$ that I find the coin (A) and you don't (B') is $w(B', A) = 1/2$, since the only alternative (you find the coin and I don't) is equally likely. Again, this is different from $w(B')w(A) = 1/4$. We say that, in this example, A and B (or A and B') are correlated (or not statistically independent).

In such a situation, the *joint* probability $w(B, A)$ of A and B can be written in the following way:

$$w(B, A) = w(B|A)w(A). \tag{A.4}$$

Here $w(B|A)$ is called the *conditional probability* of B happening given that A takes place. In the above example $w(B|A) = 0$: The conditional probability that you find the coin given that I have found it is zero. On the other hand, you not finding the coin (B') is certain if I have found it and so $w(B'|A) = 1$. We then have $w(B, A) = w(B|A)w(A) = 0 \times (1/2) = 0$ and $w(B', A) = 1 \times (1/2) = 1/2$.

In the case of statistically independent A and B the conditional probability of B becomes independent of A,

$$w(B|A) = w(B),$$

and Eq. A.3 is recovered.

In this book, there are many examples of correlated events. For example, the locations of a molecule measured at two different times are correlated provided that the time delay between the measurements is not too long.

In the above example, A (I find the coin) and B (you find the coin) are said to be *mutually exclusive*—it is impossible to have both. *If A and B are mutually exclusive then the probability that either of them happens will be the sum*

$$w(A) + w(B).$$

The two outcomes, H and T, of a single coin toss offer another example of mutually exclusive events. Thus the probability that we get either heads or tails is $w(H)+w(T) = (1/2) + (1/2) = 1$. The probabilities of all possible mutually exclusive outcomes of an experiment must add up to 1. For example, we have

$$w(HH) + w(HT) + w(TH) + w(TT) = 4 \times 1/4 = 1.$$

A.2   RANDOM NUMBERS AND THEIR DISTRIBUTIONS

Each outcome of a coin toss could be associated with a number. For example, we could play the game, in which you pay me $X(H) = \$1$ each time it is heads and I pay you $-X(T) = \$1$ each time it is tails. We call X a random number (which, in this case, takes on two possible values, $X(H)$ and $X(T)$). It is easy to calculate my expected gain in this game. If we toss the coin N times, we expect, on the average, $Nw(H)$ heads and $Nw(T)$ tails. My total gain is thus

$$G = X(T)Nw(T) + X(H)Nw(H)$$

and the average gain per coin toss is

$$G/N = X(T)w(T) + X(H)w(H) = (1/2) - (1/2) = 0.$$

It is straightforward to generalize this rule to a case where we have a random number X that can take on a set of values $X_1, X_2, \ldots X_n$ with probabilities $w_1, w_2, \ldots, w_n$. Then the average value of X is

$$\langle X \rangle = \sum_{i=1}^{n} w_i X_i.$$

Possible values of a random number X do not have to be discrete. For example, X could be the time before a light bulb burns out—clearly it is a real positive number. Then we can define a *probability density* $w(X)$ such that $w(X)dX$ is equal to the probability of finding the random number in the range between X and $X + dX$. The probability that X has any value is

$$\int_{-\infty}^{\infty} w(X)dX = 1$$

and the mean value of X is

$$\langle X \rangle = \int_{-\infty}^{\infty} Xw(X)dX.$$

If two random numbers X and Y are statistically independent, then the mean of their product is equal to the product of their means:

$$\langle XY \rangle = \langle X \rangle \langle Y \rangle.$$

Indeed, let $w_X(X)$ and $w_Y(Y)$ be their respective probability densities (the case of discrete random numbers is analogous). Then their joint probability density is $w(X, Y) = w_X(X)w_Y(Y)$ so we have

$$\langle XY \rangle = \int dX \int dY w_X(X)w_Y(Y)XY = \langle X \rangle \langle Y \rangle.$$

A.3   RANDOM WALKS

There are many examples of random walks in this book. A molecule in a gas, for example, travels along a straight path until it collides with another molecule. We can think of the resulting trajectory as a random walk consisting of discrete steps (i.e., straight segments). Starting from some initial position, which we will take to be at the coordinate origin $\mathbf{r}_0 = (x_0, y_0, z_0) = (0, 0, 0)$, the consecutive steps will be enumerated from 1 to n. The position of the walker after the n-th step is $\mathbf{r}_n = (x_n, y_n, z_n)$. The length and the direction of the i-th step are specified by the 3-dimensional vector

$$\Delta \mathbf{r}_i = (\Delta x_i, \Delta y_i, \Delta z_i) = (x_i - x_{i-1}, y_i - y_{i-1}, z_i - z_{i-1}).$$

We will assume that there is no bias in the direction of each step, i.e., $\langle \Delta \mathbf{r}_n \rangle = 0$. The location of the walker after n steps is given by

$$\mathbf{r}_n = \Delta \mathbf{r}_1 + \Delta \mathbf{r}_2 + \cdots + \Delta \mathbf{r}_n. \tag{A.5}$$

The mean position obtained through averaging over many realizations of such a random walk is zero, $\langle \mathbf{r}_n \rangle = 0$, since the mean displacement in each step is zero. This is a consequence of the position being a vector: Displacements in all directions are possible thus averaging out to zero. To quantify how far the random walker travels in n steps, one could measure the absolute distance traveled, $\langle |\mathbf{r}_n| \rangle$. It is, however, more convenient to consider the mean-square displacement, $\langle \mathbf{r}_n^2 \rangle$. Using Eq. A.5:

$$\langle \mathbf{r}_n^2 \rangle = \langle \Delta \mathbf{r}_1^2 \rangle + \langle \Delta \mathbf{r}_2^2 \rangle + \cdots + \langle \Delta \mathbf{r}_n^2 \rangle + \langle \Delta \mathbf{r}_1 \Delta \mathbf{r}_2 \rangle + \cdots.$$

Assuming statistical independence of displacements in different steps, we find

$$\langle \Delta \mathbf{r}_i \Delta \mathbf{r}_j \rangle = \langle \Delta \mathbf{r}_i \rangle \langle \Delta \mathbf{r}_j \rangle = 0, i \neq j,$$

so that only the terms containing squares of displacements survive. Finally, assuming that the mean square step length

$$l^2 = \langle \Delta \mathbf{r}_i^2 \rangle = \langle \Delta \mathbf{r}^2 \rangle \tag{A.6}$$

is independent of the step number i, we find:

$$\langle \mathbf{r}_n^2 \rangle = n \langle \Delta \mathbf{r}^2 \rangle = n l^2. \tag{A.7}$$

Therefore, the typical distance traveled by a random walker in n steps grows as the square root of the number of steps, $\sqrt{n}$.

A complete description of a random walk is expected to provide the probability distribution $w(\mathbf{r}_n)$ of the distance traveled. Our description so far appears to be incomplete since the values of $\langle \mathbf{r}_n \rangle$ and $\langle \mathbf{r}_n^2 \rangle$ we found could result from many different distributions. Moreover, the assumptions that we have made about the random walker encompass walks with diverse properties. For example, the random walker could choose, with equal probabilities, to make a step of constant length l along the x-, y-, or z-axis in either the positive or negative direction, which results in a walk on a cubic lattice. Or, the step length could be constant (and equal to l) but its direction could be completely random. In either case Eqs.A.6 and A.7 would be satisfied. Rather than being constant, the step length itself could be a random number with any probability distribution satisfying Eq. A.6. Despite their apparent diversity, all of these walks, when viewed at a sufficiently coarse scale, have the same properties. More precisely, when $n \gg 1$, the probability distribution of the vector $\mathbf{r}_n = x_n, y_n, z_n$ describing the distance traveled is given by

$$w(x_n, y_n, z_n) = w_x(x_n)w_y(y_n)w_z(z_n),$$

where each of its components has the following distribution:

$$w_x(x_n) = \left(\frac{3}{2\pi nl^2}\right)^{1/2} \exp\left(-\frac{3x_n^2}{2nl^2}\right) \tag{A.8}$$

(replace x_n by y_n or z_n for the distribution of the other two components). Eq. A.8 is a consequence of the so-called *central limit theorem*. Roughly speaking, this theorem states that the sum of a large number of statistically independent numbers has a Gaussian distribution. Specifically, Eq. A.8 results when this theorem is applied to the sum

$$x_n = \sum_{i=1}^{n} \Delta x_i$$

provided that n satisfies the condition $n \gg 1$. One must keep in mind, however, that Eq. A.8 is an approximation for any walk with a *finite* number of steps. In particular, it obviously breaks down when x_n becomes comparable to or greater than nl, since the total distance traveled by a random walker cannot possibly exceed the number of steps n multiplied by the length l of a single step.

B Elements of Statistical Mechanics

B.1 CANONICAL (GIBBS) DISTRIBUTION

It is more convenient to formulate the ideas of statistical mechanics using the quantum mechanical language, although the ensuing general relationships will not depend on whether quantum or classical mechanics is used to derive them. In quantum mechanics, any system can adopt a (finite or infinite) set of states $i = 1, 2, 3, \ldots$ with the corresponding energies $E_i = E_1, E_2, \ldots$. Multiple states can have identical energies so that the index i enumerates states rather than energy levels. The values of the energies E_i may form either a discrete set or vary in a continuous manner. Gibbs postulated that, if a system is in thermal equilibrium with a large thermal reservoir, then the probability of finding it in a state i is given by

$$w_i = e^{-\beta E_i}/q. \tag{B.1}$$

Here

$$q = \sum_i e^{-\beta E_i} \tag{B.2}$$

is called the partition function of the system and the parameter β is defined as

$$\beta = \frac{1}{k_B T}, \tag{B.3}$$

where $k_B \approx 1.38 \times 10^{-23}$ Joules/K is Boltzmann's constant and T is the temperature measured in degrees Kelvin (K). Eq. B.1 is called canonical or Gibbs distribution. When applied to ensembles of noninteracting molecules, this distribution is also known as Boltzmann's distribution. The Gibbs postulate is supported by several lines of reasoning, for example:

(1) Boltzmann showed that Eq. B.1 represents the most probable way to distribute a finite amount of energy among the available states of noninteracting molecules. To illustrate this, imagine a group of people each carrying a certain sum of money denoted M_0. Suppose now that they start exchanging money randomly using the following rule: Pick two individuals at random from the group and have the first person give the other person one dollar. If the first person has no money then the exchange does not take place. For a sufficiently large group of people and after a sufficient number of exchange transactions, the fraction of people that happen to possess M dollars will be given by

$$w(M) = e^{-M/M_0}/M_0. \tag{B.4}$$

Now replace people by molecules and call M the energy of the molecule. Then the above result can be interpreted as the Boltzmann distribution with

$\beta = 1/M_0$. The derivation of Eq. B.4 can be found in statistical mechanics texts and will not be presented here. Eq. B.4 has interesting implications for wealth distribution as it implies that random money exchanges lead to most people being predominantly poor.

(2) If we assume that, for a given system, w_i depends only on the energy of the state, E_i (i.e., $w_i = w(E_i)$, where $w(E)$ is some function) then the exponential form of this dependence, as in Eq. B.1, follows from the following two arguments. First, since energy is defined to within an arbitrary constant, an arbitrary shift in the energy of each state, $E_i \rightarrow E_i + \epsilon$, cannot lead to a change in any physical property. In particular, the ratio of the occupancies of two different states, i and j, must not change upon an energy shift:

$$w(E_i)/w(E_j) = w(E_i + \epsilon)/w(E_j + \epsilon).$$

It can be shown that, aside from a constant, the only $w(E)$ that satisfies this condition is an exponential function. Second, consider two noninteracting objects, both in thermal equilibrium with their surroundings. If we combine them into a single supersystem then the probability distribution for the states of the supersystem must satisfy the equation

$$w_{i,j} = w(E_i + E_j) = w(E_i)w(E_j).$$

This, again, leads to the conclusion that $w(E)$ must be an exponential function of energy.

B.2   THE PARTITION FUNCTION AND THE FREE ENERGY

The partition function q defined by Eq. B.2 plays a central role in statistical mechanics because other measurable properties of a material can be computed if q is known. Moreover, the *free energy* of a system, G, is related to the partition function as follows:[1]

$$G = -\frac{1}{\beta} \ln q. \tag{B.5}$$

To illustrate this connection, consider the average energy of a system,

$$\langle E \rangle = \sum_i w_i E_i = q^{-1} \sum_i e^{-\beta E_i} E_i = q^{-1}(-d/d\beta) \sum_i e^{-\beta E_i} = -d \ln q/d\beta.$$

Using Eqs.B.3 and B.5, we can rewrite this as

$$\langle E \rangle = G + TS, \tag{B.6}$$

where

$$S = -dG/dT. \tag{B.7}$$

---

[1] More precisely, Eq. B.5 defines the Helmholtz free energy of the system. It should be kept in mind that, in statistical mechanics textbooks, the symbol G is more commonly used for the Gibbs free energy. No distinction between the two kinds of free energy is, however, made in this book.

If one identifies $\langle E \rangle$ with the internal energy of the system (per molecule) and S with the entropy, then Equations B.6 and B.7 are the standard equations from thermodynamics relating internal energy, free energy, and entropy.

In many chemical applications we associate a molecular conformation with a certain subset of states. For example, the conformations A and B discussed in Chapter 3 could correspond to the protein molecule in its unfolded and folded forms (see Fig. 3.1). An unfolded protein can assume numerous shapes but even the folded protein, which has a better defined molecular geometry, may exercise excursions (such as vibrations) away from its average structure. If we think of conformations as subsets of molecular states then the partition function and the free energy corresponding to each individual conformation can be defined. For example, for the conformations A and B from Chapter 3 we have:

$$q_{A(B)} = \sum_{i \in A(B)} e^{-\beta E_i}$$

and

$$G_{A(B)} = -\frac{1}{\beta} \ln q_{A(B)}.$$

The probability that the molecule adopts a given conformation can now be written as

$$w_{A(B)} = \sum_{i \in A(B)} w_i = \frac{q_{A(B)}}{q},$$

where $q = q_A + q_B$. Therefore, relative values of the free energies corresponding to different conformations can be determined directly from their occupancies:

$$G_{A(B)} = -\frac{1}{\beta} \ln w_{A(B)} + \text{constant.} \tag{B.8}$$

These definitions can be further extended to the case where conformations are differentiated by the value of some continuous parameter λ. An example of this kind is encountered in Chapter 8, where the role of such a parameter is played by the molecular extension x. As a result, the free energy $G = G(x)$ is a continuous function of x. More generally, let $\lambda = \lambda(\mathbf{r})$ be some quantity that depends on the $3N$-dimensional vector $\mathbf{r}$ describing the configuration of the N-atomic molecule. Then the free energy, as a function of λ, is defined (to within a constant) through an extension of Eq. B.8 as

$$G(\lambda) = -\frac{1}{\beta} \ln w(\lambda),$$

where $w(\lambda) \equiv w[\lambda(\mathbf{r})]$ is the probability distribution of the parameter λ. A low-dimensional description of a molecular system provided by the free energy curve $G(\lambda)$ is of significant utility in single-molecule studies, where the choice of the parameter λ is often provided by the experimental signal itself. When λ is chosen so as to measure the progress of a chemical reaction from a reactant to a product conformation, it is often referred to as the "reaction coordinate." Thus defined, the reaction coordinate is, of course, not unique. One-dimensional free energy curves are also commonly used to describe the thermodynamics of phase transitions, where λ is chosen to be an appropriate "order parameter" that distinguishes between ordered and disordered phases of a material.

In conclusion of this section, three useful results will be pointed out. The first one is concerned with the partition function of two noninteracting systems, say 1 and 2. The total energy of the combined system composed of 1 and 2 is the sum of the individual energies. It is then obvious from Eq. B.2 that the total partition function of the combined system is the product of the individual partition functions of its components,

$$q = q_1 q_2.$$

This result will allow us to calculate the partition function of a system that contains many molecules. It also applies to the case where the total energy of the system can be decomposed into independent parts (e.g., translational energy, rotational energy, vibrational energy, etc.) so that the total partition function becomes the product of partition functions for translation, rotation, vibrations, and so forth.

The second result is the classical definition of partition function and classical calculation of Boltzmann-averaged quantities. The state of a one-dimensional classical particle of mass m can be specified by the values of its position x and momentum p, while its energy is

$$E(x, p) = \frac{p^2}{2m} + V(x),$$

where $V(x)$ is the potential energy. It is natural to expect that the summation over the index i in Eq. B.2 should become integration over x and p when classical mechanics is used:

$$q = \int dx dp\, e^{-\beta E(x,p)}.$$

The quantum partition function described by Eq. B.2 is, however, a dimensionless quantity while the above integral is not. The correct classical limit of Eq. B.2 turns out to be

$$q = \int \frac{dx dp}{2\pi \hbar} e^{-\beta E(x,p)}, \tag{B.9}$$

where

$$\hbar = \frac{h}{2\pi} \approx 1.05 \times 10^{-34} J \times s$$

is Planck's constant (and where the quantity $h = 2\pi\hbar$ is also commonly referred to as Planck's constant). While a rigorous proof of Eq. B.9 will not be pursued here, it can be physically motivated by considering the quantum mechanical uncertainty principle, which states that the momentum p and the coordinate x of a quantum particle cannot be measured simultaneously and that the uncertainties Δp and Δx in measuring these quantities are related by

$$\Delta x \Delta p \approx 2\pi \hbar.$$

If we represent classical "states" as points in the (x, p) plane, then, in view of the uncertainty principle, such states cannot be resolved in quantum mechanics. Rather, quantum states can be thought of as "cells" of the (x, p) plane, each with an area of

$2\pi\hbar$. The division by $2\pi\hbar$ in Eq. B.9 takes care of proper counting of such cells. The general rule for going from quantum summation to classical integration is, therefore, as follows:

$$\sum_i \cdots \rightarrow \int \frac{dxdp}{2\pi\hbar} \cdots$$

If $f(x, p)$ is some function of the classical coordinate and momentum then its canonical average can be computed, classically, as

$$\langle f \rangle = \int dxdp w(x, p) f(x, p) = q^{-1} \int \frac{dxdp}{2\pi\hbar} f(x, p) e^{-\beta E(x,p)}.$$

These results are easily generalized to include multidimensional systems. In particular, for a system of N classical particles with the coordinates $\mathbf{x} = (x_1, y_1, y_2, x_2, y_2, z_2, \ldots, x_N, y_N, z_N)$ and corresponding momenta $\mathbf{p} = (p_{x1}, p_{y1}, p_{z1}, p_{x2}, p_{y2}, p_{z2}, \ldots, p_{xN}, p_{yN}, p_{zN})$, the classical partition function is given by a $3N$-dimensional generalization of Eq. B.9,

$$q = \int \frac{d^{3N}\mathbf{x} d^{3N}\mathbf{p}}{(2\pi\hbar)^{3N}} e^{-\beta E(\mathbf{x},\mathbf{p})}.$$

Finally, the third useful result is the partition of a harmonic oscillator, whose energy is given by

$$E(x, p) = \frac{p^2}{2m} + \frac{\gamma x^2}{2} = \frac{p^2}{2m} + \frac{m\omega^2 x^2}{2},$$

where γ is the spring constant and $\omega = \sqrt{\gamma/m}$ the oscillator frequency. Substituting this energy into Eq. B.9 and performing integration, one finds the classical partition function,

$$q_{cl} = \frac{1}{\beta\hbar\omega} = \frac{k_B T}{\hbar\omega}.$$

This result is, however, an approximation. The exact quantum mechanical result is obtained by substituting the exact expression for the quantum energy labels of the oscillator given by

$$E_i = \hbar\omega \left(i + \frac{1}{2} \right), i = 0, 1, \ldots,$$

into Eq. B.2. Summing the resulting geometric progression, one finds

$$q = \frac{e^{\beta\hbar\omega/2}}{e^{\beta\hbar\omega} - 1} = \frac{1}{2\sinh(\beta\hbar\omega/2)},$$

which reduces to its classical limit q_{cl} when the thermal energy is much greater than the spacing between the oscillator's energy levels, i.e.., $k_B T \gg \hbar\omega$.

B.3   MAXWELL-BOLTZMANN DISTRIBUTION AND THE EQUIPARTITION THEOREM

The probability distribution of an atom's velocity components (v_x, v_y, v_z) provides an example of Eq. B.1:

$$w(v_x, v_y, v_z) = q^{-1} \exp \left(-\frac{mv_x^2/2 + mv_y^2/2 + mv_z^2/2}{k_B T} \right) = w_x(v_x)w_y(v_y)w_x(v_z),$$

(B.10)

where m is the atom's mass. The normalization constant q is given by the product $q = q_x q_y q_z = q_x^3$, where

$$q_x = \int_{-\infty}^{\infty} dv_x \exp \left(-\frac{mv_x^2}{2k_B T} \right) = (2\pi k_B T/m)^{1/2}.$$

Eq. B.10 is known as the Maxwell-Boltzmann distribution. The average kinetic energy of the atom,

$$\langle mv_x^2/2 + mv_y^2 + mv_z^2/2 \rangle = \langle mv_x^2/2 \rangle + \langle mv_y^2/2 \rangle + \langle mv_z^2/2 \rangle = 3\langle mv_x^2/2 \rangle,$$

can now be estimated from

$$\langle mv_x^2/2 \rangle = \int_{-\infty}^{\infty} dv_x w(v_x) \exp \left(-\frac{mv_x^2}{2k_B T} \right) = k_B T/2$$

so that the total average energy is $3k_B T/2$. The result does not depend on the mass of the atom. In fact, it solely relies on the quadratic velocity dependence of the kinetic energy while the proportionality coefficient (i.e., $m/2$) is unimportant. By the same token, for example, the mean potential energy of a one-dimensional harmonic oscillator, whose microscopic value is given by $\gamma x^2/2$ with γ being a spring constant and x an appropriate displacement, must be independent of γ and equal to $k_B T/2$. The total average oscillator energy is then $k_B T/2 + k_B T/2 = k_B T$. These observations are consequences of the more general *equipartition theorem*, which states that the average contribution corresponding to each independent quadratic term in the molecule's energy (such as $mv_x^2/2$ or $\gamma x^2/2$) is the same and equal to $k_B T/2$.

Index